EVOLUTION IN A NUTSHELL

by Martin Malloy

2ND EDITION

Evolution in a Nutshell
by Martin Malloy
(2nd edition)

Copyright © 2011

ISBN 978-1-257-65201-3

PROLOGUE

Evolution in a Nutshell examines the peculiarities of the human primate and presents speculative reasons for changes in the adaptive social structure of the species. Further, since concepts are important in evaluating evolutionary phenomena they are developed from the perspective of natural selection favoring traits with their tagalong genes that are best suited for the survival and reproduction of a species.

Concepts, for example, formed by biological scientists studying fossils and other data convinced them that evolution is deterministic, but without purpose. The scientific conclusion is illustrated by common and uncommon animals such as elephants and duckbill platypuses or eagles and tapeworms. The characteristics of the acutely different animals are explained by evolution acting on a potpourri of genes produced by the random process of genetic variation. In a nutshell of evolution, nothing is planned.

Concepts function to visualize the reasons emerging organisms radiate or spread out and adapt their physical and behavioral characteristics to new niches. Concepts help people understand why large dinosaurs were cold blooded and large mammoths warm blooded. Conceptualizing makes it clearer why the two big animals became extinct and are represented today by smaller descendants. Indeed, evolutionary concepts make it easy to infer worms cannot act with comprehension while chimpanzees plan for the future. Seeing the larger picture helps one to envision the physical appearance of the common ancestor of orangutans, gorillas, chimpanzees and humans. A rose by any other name would smell as sweet is a concept as much as a man or woman by any other name are still Great Apes.

There are several terms that are essential in conceptualizing evolution from its beginning, and specifically, human evolution. These are: adaptation, common descent, DNA, mutation, natural selection, nonadaptive, speciation, species, pseudospeciation and egalitarianism. They are not difficult terms to develop into concepts and the two that strongly apply to human evolution, and the current dilemma of human conflict are pseudospeciation and egalitarianism. These terms are in the glossary which might be examined as a prelude to reading Evolution in a Nutshell.

Evolutionary concepts are good for speculation, and for forming hypotheses and theories. Concepts are akin to intuition, and hunches, or that 'gut feeling', but as nothing is sure, they don't offer yes or no answers. Scientists can't say yes or no that viruses evolved before bacteria or Neanderthals evolved from Homo erectus. However, they can say viruses and bacteria coexisted in the Precambrian seas and Neanderthals had larger brains than humans. Although more is known about Neanderthals than any other human relative, no one can say Neanderthals were smarter or less intelligent than people or what caused them to become extinct. Neanderthals buried their departed in tended graves and fossils of flower seeds were found with the remains. Flowers are sentimental symbols of love and compassion and other qualities that sustain an egalitarian social organization.

An anthropological view of love and compassion is they are adaptive properties which are manifested in humans and chimpanzees. The discovery of sentimental symbols in Neanderthal society infers the properties are innate in some or all primates. There is also the negative side of humans and chimpanzees in their hatred and aggressive methods of killing their fellows. Intuition, that old gut feeling, says something does not fit like the square peg in a round hole. The human primate appears to have too many square pegs for the number of square holes. *Evolution in a Nutshell* examines the uncommon mode of human evolution that might have produced too many square pegs for the nearly hairless primate. The social organization typical of the Western World and the pristine society of human hunter–gatherers are compared to the social structures of disturbed and undisturbed chimpanzees.

A few grammatical liberties are taken in the book with words and phrases. Types and similes are expressed by such as and like which writers use so often the words become redundant. Therefore, I decided to be guided by the sound, or how they fit in a sentence. Commas were placed or omitted before and after quotes, not as much by grammatical rules as by the smooth flow they tender traveling eyes.

That is, words and phrases were employed to trip gently within the ear instead of on the tongue while the commas act as ocular balms. Some words were coined to better describe phenomena for which existing words are inadequate. After all, John Milton coined pandemonium to portray the chaos a gang of rebellious angels made for themselves. Pandemonium is still a good word for the continuing confusion that economic systems of direct competition have inflicted on human societies.

Science writing is usually dry so throughout the pages I sprinkled a little humor and satire for lighter reading and tried not to be corny or detract from the seriousness of the subject. Genetics and a few other technical subjects addressed by *Evolution in a Nutshell* are simplified and discussed mainly in the first few chapters. The technical material is important in understanding evolutionary events and to seeing evolution as a panorama of random occurrences.

The book is best read chapter by chapter because it is not an anthology of essays, but a continuum of information and emerging concepts. The book presents a chronological account of evolution from the creation of life by natural selection in the Precambrian Period until humans emerged. However, the account is not meant to indicate a planned or purposeful progression.

Moreover, faith based speculations are not considered except for historical reference. Passages from religious texts are cited such as those by stern prophets urging social conformance or kindly behavior advocated by humanitarian preachers. Quotes from other literary sources are used to flavor the material along with occasional satire.

Evolution in a Nutshell does not embody optimism or a 'things will get better' prophesy. The theme is objective and suggests a practical, but disquieting solution to the recurring dilemmas of inherently loving and compassionate humans. The book is not meant to be a downer, but purposeless evolution isn't compatible with 'hope springs eternal'.

The many things that exist in the world which offend human sensibilities are considered from an evolutionary prospective. Warfare, crime, murder, suicide, prejudicial judgments and the countless stressful situations such as "irreconcilable differences" between married couples exemplify these offences. Also addressed by evolutionary hypotheses are curiosities specific to humans whose causes are unanswered, which include homosexuality, puny testicles and menopause. The theme is meant to be a seed which will grow and be cultivated by a few who read *Evolution in a Nutshell.*

ACKNOWLEDGMENTS

Many thanks to my family and friends who listened to me all these years talking about our nonadaptive social organization and uncommon mode of evolution from an ape. Their interest and feedback gave me more ideas for developing my manuscript. My wife, Elayne, is a natural born listener with whom I have had countless discussions since I began thinking about writing a treatise on evolution. Her help in editing was invaluable. My thanks to my friend Michael Rich who formatted *Evolution in a Nutshell* for publication.

The germinal thought for *Evolution in a Nutshell* began many years ago when I was a child in a small town in rural Georgia where I was born and raised during the Great Depression. The economy was poor and only a few people were well off. Most worked hard for what little they had and others endured a life of poverty. Everyone knew each other in that agricultural community in those different times. People were yet to be polarized, one result of a prospering economy.

My father was a physician, a Tulane graduate, and a man of infinite compassion who was venerated for his competence, dedication and loyalty. When he died there wasn't enough room in the church for the people at his funeral. The minister, organist, choir and congregation all cried. Those were days of simplicity and innocence that have all but vanished save in memories.

My father made house calls to attend the sick and sometimes I would ride with him and wait in the car or sit on the porch or visit with the family. One day I impulsively said to him, "daddy do you ever feel sorry for all the little children who don't have anything?" I don't remember if we were in his car or at home, but I will never forget the occasion. He looked away for a moment and seemed to search for something far into the distance. Softly, and now I know, sadly, he said "constantly." I will always remember the look in his eyes which I saw three more times during my life with him.

I don't know why I asked him that question unless it was a curious response to my feeling sorry for unfortunate people and I wanted his consolation. One thing is certain, in the back of my young mind I wondered why children had to suffer. Perhaps I wondered why anyone had to suffer. Many years later I concluded from my studies and observations that we are born with the quality of compassion. We all feel pity for the suffering of others. Only people who are disturbed for some reason are exempt from those emotions. Those people may be indifferent and even cruel. There were many social situations I observed in the little town that contributed to *Evolution in a Nutshell*, but most of all I thank my daddy for his answer long ago.

CHAPTER 1

Planet of the Apes is a sci-fi movie about apes taking over the world after a nuclear holocaust. The story takes place in 3978 AD and deplores the foolishness of people for not preventing the disaster. The film is based on a 1963 doomsday novel published a year after the Cuban missile crisis that brought the United States and former Soviet Union close to war. Released in 1968 at the height of the Vietnam conflict, the film's appeal was enhanced by some timely discoveries in the 1960s.

One eye-opener was Jane Goodall's observation of chimpanzees in their African habitat making tools, which was thought a skill practiced only by humans. Laboratory chimps were solving challenging problems and learning to 'talk' with symbols. Although disturbing to some people, human and chimpanzee DNA analysis revealed the two primates are more closely related to each other than to any other species. The 1960s saw progress in human rights and the way people treated their hairy kindred. Chimpanzees raised in human families showed they were sentient and intelligent beings sharing the same emotions with their larger brain relatives.

Intelligent man

$E = MC^2$

Intelligent chimp

1 + 1 bananas = 2 bananas

Another revelation of chimpanzee intelligence was a chimp named Ham led America in the race for space. The country had been surprised and embarrassed in 1957 by the sensational launch of the Russian satellite Sputnik. While Sputnik impressively orbited the earth, the United States resolved to prove the Western World was superior to its communist adversaries. The Astrochimp Ham came to the rescue as the vanguard of the space program that put an American on the moon in 1969.

The intrepid cosmonaut made history in 1961 when he blasted off toward the stars from Cape Canaveral, Florida in a Mercury Redstone rocket. Ham was a real live son of Uncle Sam who returned data from his spaceship to the command center. The hairy hero paved the way for the first American to orbit the earth. Truly, the decade of the 1960s was filled with surprises.

Ham and a cousin

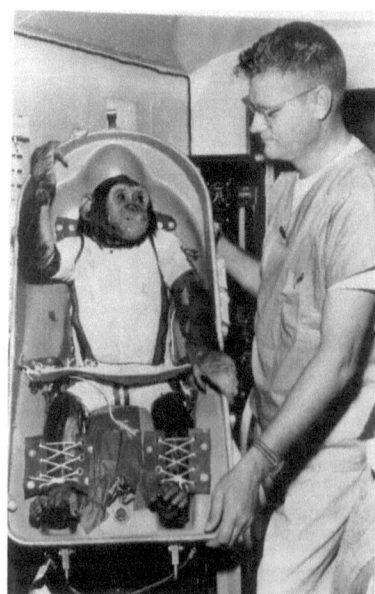

Some of Ham's 'descendants' in *Planet of the Apes* were
scientists who dissected human brains in medical
experiments. After all, turn about is fair play. The futuristic
chimpanzees had reversed evolution and adapted traits in two
thousand years that took humans five million years of
evolution to obtain. The apes walked upright, had dexterous
hands, spoke good English and possessed human intelligence.

Conversely, people had degenerated into speechless vassals
of the ruling ape society. Their degeneracy was an allegorical
assault on the notion of human superiority. The apes were
fervent fundamentalists, religious extremists who were guided
by their Sacred Scrolls, a simian version of the Old
Testament. Their holy book cautioned them to beware of
humans for they are "the Devil's pawns."

Two and a half million years earlier, the first species of their
relatives with larger brains began to evolve and eventually
reduced all apes to second-class primates. Every dog has its
day, and when the futuristic apes took over it was payback
time. The revolutionary apes subjugated people like the
revolutionary people of France overthrew their monarchy in
1793. The apes did not establish a one for all and all for one
egalitarian society. Instead, they maintained a mildly ranked
society with some features of the human societies that were
driving them to extinction until the nuclear holocaust.

The ape society of 3978 AD was not repressive nor was it
entirely classless, but a middle of the road organization like
lukewarm water between hot and cold. Their centrist
arrangement was not an egalitarian system or a socialistic
Utopia. Perhaps the apes favored egalitarianism at first, but
that preference was trashed by their elevated intelligence,
like people trashed their adaptive society after the
Agricultural Revolution. Regardless, the apes did not
construct the principles of their society to assure an

unranked organization. Their society was hierarchal with social norms to follow and laws to obey.

They had disagreements that could be civilly resolved, except in matters of state religion, which was strict fundamentalism. The apes were expected to follow and not question the tenets of their required creed. Unless a society is completely egalitarian, which theirs wasn't, it's credible hierarchies of different control levels will form mechanically. The development of their social system was directed by their novel and superior intelligence that was humanlike. Considering religious zealots controlled their society like extremists control some human fundamentalist sects, those future apes could have been setting themselves up for World War IV.

Although their social system wasn't rigidly tiered, it was a brand of socialism divided by employment categories. Chimpanzees were scientists and intellects, orangutans the authoritative officials and bureaucrats. Gorillas were crude soldiers portraying their 1960s stereotype of aggressive animals. The negative stereotype of gorillas was dispelled in the 1980s when Koko, a home-raised female gorilla showed gorillas were gentle apes capable of love and compassion.

The futuristic ape society was running smoothly when an astronaut from 1972 crash-landed on earth like a fly landing in soup. The astronaut had blasted off into outer space before the nuclear war. His spaceship was sucked into a time warp that thrust him ahead two thousand years. When the craft landed back on earth the confused fellow thought he was on another planet. The apes captured the weird man who could speak and planned to castrate and lobotomize the freak for their primate studies.

However, two chimpanzee scientists befriended the astronaut and helped him escape. Their good deed didn't go unpunished and turned out to be tragic for both the apes and humans. The astronaut went to a region where mutant descendants of the holocaust survivors were living. The disfigured mutants lived underground and worshiped a doomsday bomb. While the bumbling astronaut was there he accidentally caused the bomb to explode and the world was destroyed.

Indeed, the serpent in the garden was the interloping astronaut who justified the passage in the Sacred Scrolls that Homo sapiens are the Devil's pawns. The great paradox is that people have an infinite capacity for love in contrast to a checkered history of warfare. Provided love is a quality Nature adapted for a purpose, it is credible that aggression is a frustrated response to circumstances that people aren't adapted to handle. Whatever the reason, hopefully the mystery will be clarified by conclusions drawn from theories of human evolution.

The prevailing theory of human evolution proposes hominids, or humans and their extinct relatives evolved from an African ape five million years ago at the beginning of the Pliocene Epoch. The ape was the common ancestor of hominids and the great apes. The great apes evolved before hominids in the order: orangutans, gorillas and chimpanzees. Many scientists suggest classifying extinct hominids and extant Homo sapiens as great apes. The web site of the Smithsonian Hall of Human Ancestors currently exhibits recognized hominid species with explanatory texts and is recommended to view.

Ancestral ape

Some human ancestors and relatives

Ancestral bipedal ape   Australopithecus africanus

Homo heidelebergensis   Homo nenderthalensis

Alpha to Omega

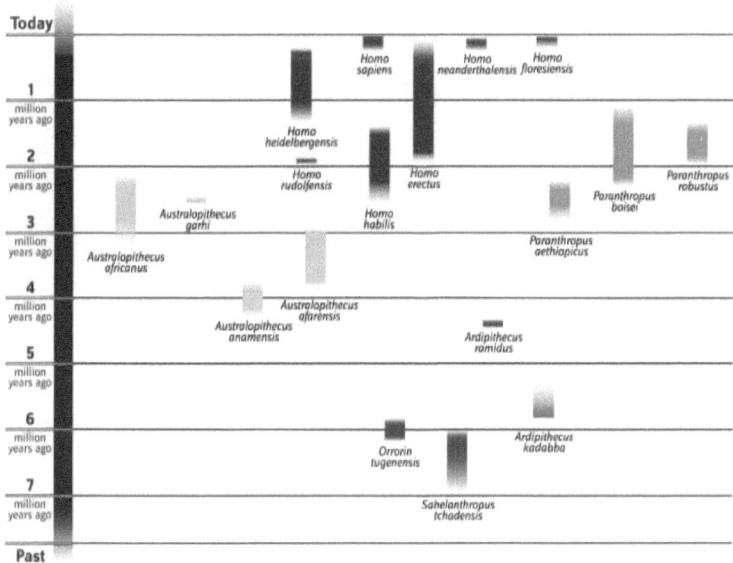

The first hominids to evolve, according to theory, descended from tree apes, which were slowly losing their forest habitats during prolonged climate changes. The imperiled apes could emigrate easier and forage on the African plains by bipedal or erect walking. The apes knuckled walked on all fours using the knuckles on their hands for support. They needed to modify their bodies to improve their survival and reproductive prospects in the new environment.

The apes began spending more time on the ground as their adaptive environs declined. Nature gradually altered their bodies to stand up straight as they moved among sustaining clusters of trees. The spine curved, hips flared, legs grew longer and the toes aligned side by side. The opening for the spinal cord, the foramen magnum, was located rearward at the base of the skull. The opening began moving toward the middle, which rotated the head directly above the center of gravity causing the apes to stand erect.

Cranial and other circulatory vessels reorganized to facilitate the vertical flow of blood against the pull of gravity. A later modification of the vessels would help cool the brain of some apish émigrés on the sun-drenched plains. The upright stance of the apes exposed less body surface to solar radiation and reduced heat absorption. Standing tall on the African landscape, the ambulating primates could look for food sources and watch for danger.

Moving to the ground had its advantages, but there were huge cats and other fearsome predators roaming the plains. Fortunately, for the ape-like bipeds, they had not co-evolved with the marauding animals and were not their natural prey. However, the hominids were meaty enough to be more than a between meals snack and needed to defend themselves. The newly arrived migrants threw rocks and flailed limbs to repulse the investigative beasts and retain their stranger status. One conjecture applying to current times is the genes directing their barrages are the same--genes directive in games such as baseball and cricket.

Their defensive strategy is inferred by the intelligence and behavior of chimpanzees. Goodall says in her comprehensive book, *The Chimpanzees of Gombe*, that a chimp threw a rock "estimated to weigh 5 kilograms" (11 lbs.) and another slammed a stick with impressive force and speed. The modified apes stood together like a Roman legion and five million years later some of their descendants would pen "e pluribus unum."

Some evolutionists think bipeds arose quickly from the knuckle walking apes because of the scarcity of hominid fossils between four and five million years ago. Others believe there are undiscovered fossils that would systematize the hominids and show they emerged slowly from the stooped primates. An incomplete fossil record and the fact most fossils are teeth and bone fragments makes it difficult to categorize the extinct hominids.

Many evolutionary scientists have personal time models based on their own theories for the emergence and extinction of the hominid species. Anyone can have a personal theory of human origin that could be correct since none has been proven. Evolutionary biologist Ernst Mayr chose his theory "among the numerous interpretations that seemed to me the most likely correct one." Many questions remain in classifying the hominids, determining the lines of descent and the time of divergence. The question that has been answered to the satisfaction of evolutionary scientists is hominids evolved from apes.

When hominid fossils are excavated they are dated and the species identified or a new species is recognized. Species is a singular or plural noun with several definitions that are usually treated as a concept. One definition of a species is a population, or populations of interbreeding organisms that reproduce offspring that are 100% fertile. The genetic composition of individuals of a species is so near they are easily assigned to their species by genetic analysis. Sibling species and subspecies designate degrees of genetic kinship.

Members of a species usually look alike such as African and Asian elephants that are close relatives, but different species. The elephants are fairly easy to distinguish, which isn't always true of other related species. Offspring of related species may be fertile or sterile, or partially fertile, depending on the parent's genetic nearness. A mule, for example, is an offspring of a horse and donkey and a sterile hybrid whereas a liger is the fertile offspring of a lion and tiger. Genetic theory holds the horse and donkey are further removed from their common ancestor than the lion and tiger.

The consensus is modern humans and past hominids have a similar genetic makeup. Accordingly, their genetic similarity decreases with each species as they regress in time. Deductively, people would be closer kin to hominids living one million years ago than those living four million years ago. Knowing the specific genetic composition of each succeeding hominid would give an assessment of their degree of kinship. The same could be said of any organism and its progenitors such as elephants and extinct mammoths.

Dating and classifying their fossils gives an estimate of the timeline of an animal's descent and divergence. There are many dating methods and some are precise and others that are used for difficult situations are approximations. One means of dating is the potassium–argon method, which calculates the age of the oldest fossils. Radioactive potassium deteriorates into the gas argon, and if the amount of argon is known, the time it took the gas to form can be computed. The age of fossils less than fifty thousand years old is determined by carbon dating.

Some methods date time spans too old for carbon dating or too recent for the potassium method. Fossils of elephants, pigs, rodents and other animals with known histories are used as time markers to indirectly date fossils of unknown origin. Indirect, or relative methods are utilized when radioactive material is lacking and to evaluate the accuracy of highly technical procedures. Mayr says "the sequence of accurately dated fossils has documented evolution in the most convincing manner."

Two dated hominid genera are the gracial and robust Australopithecines and the Homo lineage. Australopithecine means southern ape that is sometimes abbreviated Australopith. The singular genus and plural genera designate a general category of closely related organisms above species. Species is the narrowest category an organism is usually assigned. Genus and species classify organisms in the binomial system devised in 1751 by Carolus Linnaeus, a Swedish naturalist.

Modern humans are classified as Homo sapiens, from Latin that was generously translated to mean sensible humans. The genus of the house cat is Felis and the species is catus, or Felis catus. A higher category for humans is the order primates and for cats it is the order Carnivora. People, cats, lions, tigers and elephants are mammals, an inclusive class. Fish, amphibians, reptiles and mammals are classes of animals included in the larger category of the Animal Kingdom, information contained in a high school biology text.

The oldest Australopith fossils are more than four million years old, but could change quickly with a new discovery. Several species of the Australopith genus such as the gracile Australopithecus Afarensis and Australopithecus Africanus existed in Africa for 3.4 million years. The robust Australopiths adapted big muscular jaws and large molar teeth to eat tough vegetation like tubers and roots. Australopiths were about the size of chimpanzees and classifying their fossils can be contentious. Some scientists believe Australopiths living at the same time were essentially varieties of the same species. Varieties often refer to closely related subspecies or sibling species.

Robust Australopithecine 2,000,000 BCE

First hominid vegan?

The Australopith bodies changed little during their earthly term and their brains remained chimp-size. The dissimilarities among the Australopith genera are deemed minor, which some evolutionists attribute to adaptions to different environments, as are human 'races'. Their chimpish bodies and brains might have spared them the troubles Homo sapiens have with their frail bodies and excessive intelligence.

Presumed footprints of two or three Australopiths were discovered in Africa by the late paleoanthropologist Mary Leakey in 1978. The prints were uncovered from 3.7 million-year old volcanic ash and were dated indirectly by the age of the ash. The prints look like tracks made by people walking barefoot in the sand. The depth and distance between the prints indicate one biped was almost four feet tall and the other slightly smaller. The primal hominids ambling along the ashen trail were about the size of two chimps.

Footprints in the Ash

Circa 3.7 million years ago

Graphic evidence of the Australopiths was the 1974
discovery of a partial female skeleton in Africa by
paleoanthropologist Donald Johanson and his colleagues. The
3.2 million year–old hominid was dated by the age of the
layer of earth in which she fossilized. Johanson classified her
as a new species, Australopithecus afarensis. She was
nicknamed Lucy after a popular song by the Beatles and was
the most significant find until that time. Lucy is often
described as a bipedal ape and basically a chimp on human
legs. Her fossil skeleton has been exhibited in several
countries.

The female hominid was three and a half feet tall and
weighed sixty–five pounds. Her face and head resembled a
chimp's face and head and her arms were longer than her
legs, a characteristic of apes. Lucy's thick bones and sinews
indicate the ape–like biped was much stronger than a person.
Many researchers think she used her arms and curled fingers

to climb trees. They say her species lived in trees like chimps where they fed and made night nests. Others believe they were mainly terrestrial and climbed trees to feed or avoid danger.

Lucy
Grandma or Auntie?

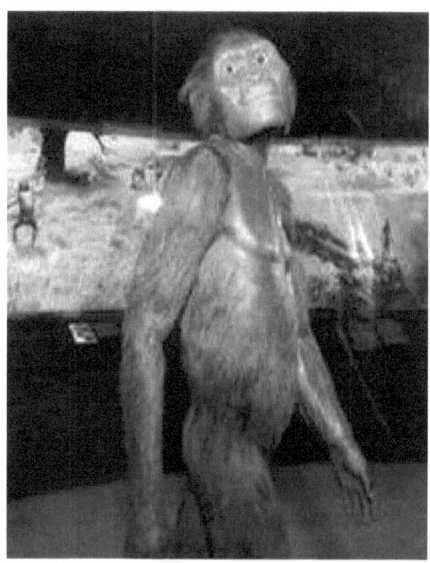

Lucy's much younger relative, Homo sapiens, sprouted on the last twig of the Homo line's family tree. The peculiar species emerged with an enigmatic large brain out of proportion with the brain of the great apes. The Homo hominids represent several species that are challenging to categorize such as Homo habilis, hypothetically the first to evolve. The oldest Habilis fossils are two and half million years old and show these early hominids had chimp-size bodies and slightly larger brains. Later fossils of Homo hominids disclose they were becoming more like humans with brain areas that were expanding unevenly.

Homo erectus, another character in the hominid picture appeared about two million years ago. Homo erectus means upright man and is well represented by fossils that show the species was physically similar to Homo sapiens. One difference is the Erectus brain was one third smaller than the brain of Homo sapiens, but like the brain shape of the Homo lineage, it's not as rounded as Australopith or ape brains. Erectus is the direct precursor or 'parent' of Homo sapiens in one model of human descent and in another model Erectus has no descendants.

Homo erectus dark skin African hominid

Let's eat!

Circa 1.6 million years ago

Evidence from ancient encampments of Homo erectus indicates the extinct species lived in hunter-gatherer bands at the time Homo sapiens evolved. Provided their social structure was the same as the peaceful hunter-gatherer tribes that have all but vanished, suggests they were nonviolent egalitarians. Artifacts of warfare and other evidence would have to be discovered to conclude they killed each other, as do modern humans. Homo erectus was so structurally similar to people that if they were dressed in modern clothing they would resemble a band of rock musicians.

Some Homo erectus tribes left Africa a million years ago and spread into many regions of the world. A theory poses Erectus gave rise to Cro-Magnon, an early Homo sapiens, fictionalized by Jene Auel's novel, *Clan of the Cave Bear*. Homo erectus was long believed to be the first hominid to leave Africa. However, fossils of a smaller skull found in Dmanisi, Georgia in southern Europe suggest an earlier hominid was the first emigrant. The smaller brain hominid was christened Homo georgicus and arrived in Europe eight hundred thousand years before Homo erectus.

Republic of Georgia

Homo georgicus. Lighter skin European

Circa 1.8 million years ago

Papa and Mama
Human ancestor?

A nearly complete skeleton of a Homo erectus boy was
unearthed in 1984 by paleoanthropologist Richard Leakey
and his associates near Lake Turkana in Africa. They named
the 1.6 million-year old specimen the Turkana boy. The
Homo lad was similar to modern humans, but the brain case
was one third smaller. Some scientists classify the adolescent
boy as another species, Homo ergaster. However, others say
there isn't enough difference to distinguish Ergaster from
Erectus and they are the same species.

Turkana Boy's skeleton

An untimely death

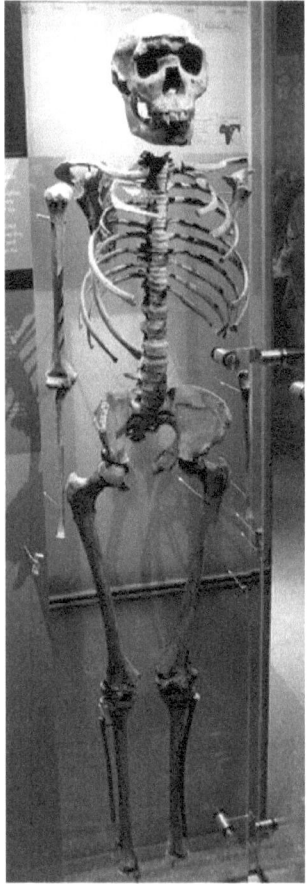

Close human relative

The skeleton of the Erectus boy shows the species was indisputably distinct from the Australopith hominids, the theoretical precursors of the Homo lineage. A proposal for the direct precursor of the Homo hominids is Australopithecus garhi, dated 2.5 million years ago. Garhi was at the 'halfway' mark between the first Australopiths and Homo sapiens and is the 'missing link', a nifty term, but unscientific and trivial. A lesser theory of Homo hominid descent is the Homo lineage evolved directly from the ancestral ape or from an undiscovered species, and not from an Australopith.

Australopithecus garhi
Circa 2.5 million years ago

Another Grandparent?

CHAPTER 2

Whatever unplanned paths the hominids took, events leading to their arrival occurred long before their ancestors left the trees. The first event took place in the early Precambrian Era. Scientists theorize the earth was hot and had recently solidified from a mass of drifting gasses. The atmosphere was a gaseous mixture, mainly carbon dioxide, ammonia, methane, hydrogen sulfide and steam. Eventually, the temperature decreased and the steam condensed into oceans and inland waters. The air and water composed a chemical 'soup' that contained the fundamental materials of living matter, including the makings of people.

The Russian biochemist Aleksandr Oparin hypothesized the materials were inorganic or nonliving substances that progressed by chance into organic or living substances. Oparin thought that lightning striking through the atmosphere into the soup transfigured the substances into elementary living matter. His ideas were supported by laboratory experiments using gases that simulated the early Precambrian atmosphere. Electricity was passed through the gasses suspended in a chamber of water. Later a chemical analysis showed the water contained amino acids, the recipes of life that build proteins. Billions of years later the recipes and random chance would concoct men, women and children.

Generating life isn't a new idea and was fictionalized by the English writer, Mary Shelly. In her 1818 novel, *Frankenstein*, a man was assembled from dead body parts and brought to life. The man accidentally got the brain of a criminal and fled the scene after murdering the chap who procured the substandard brain. Frankenstein's tale was made into horror movies in which life was infused into the piecemeal man by juicing electricity through knobs on his neck. Ironically, Mary

Shelley's story thrives today while the splendid poems of her renowned husband, Percy Shelly, are read mostly in English classes.

Frankenstein Monster

Homo horrobilis

Electrical stimuli in the Precambrian waters moved the altering substances through obscure steps from the nonliving to the living. The living system that arose in the early Precambrian is defined as the point where natural selection began to act. Darwin's Theory of evolution by natural selection proposes that organisms with traits best suited for their surroundings have the most potential to survive and reproduce. Moreover, natural selection applied to inanimate objects holds the most successful ones persist and replicate. Though natural selection is not the only means of evolution, it is the prevailing mode and an essential concept.

The simplest action of natural selection in the Precambrian soup was an emerging unit of life reproducing itself and repeating the process. Francis Crick, co-discoverer of the structure of the DNA molecule, thought the unit was RNA, the first nucleic acid. RNA differs in one element from its famed 'sibling,' DNA, popularized in crime stories to identify lawbreakers. The RNA and DNA molecules are called nucleotides and have two basic functions: to pass on hereditary characteristics from one generation to the next, and to trigger the manufacture of specific proteins that build organisms.

Crick says the duplication of RNA in the soup was "closely followed by a simple form of protein synthesis." RNA inadvertently started the life processes in the ancient seas and in time gave rise to DNA. Billions of years later the DNA of a man and woman would be incorporated in a sperm and ovum that would unite to build a baby boy or girl. The babies would grow into teenagers who would read about DNA, genes and chromosomes in their biology classes.

Genes are series of nucleotides that are stored on packaging units called chromosomes. The genotype is the total number of an organism's genes and the phenotype is its observable features. The human genotype, for instance, is the total number of a person's genes, which are composed of DNA, and the phenotype is the human body. The genetic code of all organisms consists of nucleotides that direct the reproduction of offspring that belong to a definite species. Tiger sharks give birth to tiger sharks, green frogs to green frogs and Klondike bears to Klondike bears.

Human DNA is a recipe 'written' in a molecule that directs the fertilized egg as it grows through embryonic stages into a baby boy or girl. The fertilized egg is a fertilized ovum, also called a zygote, terms that are often used interchangeably. Zygote usually refers to the ball of dividing cells between the fertilized egg and the embryo, for what it's worth. A North Carolina senator prophetically proclaimed in 1986 that life begins at conception when the sperm and egg unite, for what that's worth.

DNA Molecule Model

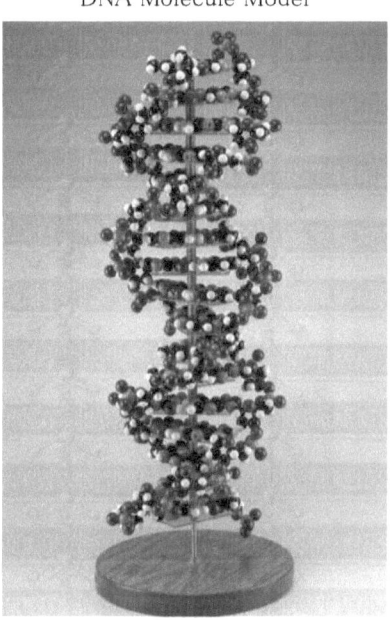

Recipe for an organism

Life began in the Precambrian seas with viruses or bacteria, but which came first is unknown. Viruses are parasitic and smaller than bacteria. Viruses, unlike other organisms don't have all the life processes such as ingestion, elimination, respiration, reproduction and growth. Viruses coming first could appear to be an intended progression from simple to complex organisms, a concept in keeping with human experience.

However, scientific consensus is evolution and its events are haphazard, and like the chicken and egg it's undetermined which came first, viruses or bacteria. Viruses don't fit the definition of true life forms since they can remain inanimate crystals indefinitely. Since bacteria were the first organisms with all the life processes, they are usually considered the prime human ancestor and not the crystallizing viruses. Viruses invade a living cell to reproduce, which is often fatal to the host. After entering a host cell in their zombie-like state, they begin to function like a living organism.

Virus

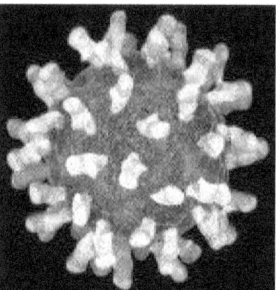

A human cousin??

Irishman Bram Stoker in another imaginative novel, *Dracula*, crossed the border between the living and nonliving. In the spine tingler Stoker published in 1897, Count Dracula of Transylvania remained lifeless in his coffin by day. At night he came to life and crept out to turn into a big bat that sucked the blood of human victims. The mystery of the cloaked Count Dracula has increased in popularity since its initial publication.

The mystery of viruses as the origin of life remains as cloaked as Dracula since they left no Precambrian fossils. One theory poses viruses were degenerate components of cellular organisms and another is viruses arose from parasites within cellular organisms. Proving they were degenerate components or came from parasites would mean they didn't seed the tree of life. However, to prove they seeded the six biological kingdoms would make them the primary human ancestor. This would denote that the rudiments of people traveled on a road paved by viruses.

There are no riddles in dating bacteria since their fossils were found in 3.5 billion year–old Precambrian strata. The early atmosphere of the Precambrian contained little or no oxygen according to scientific research. Deductively, the first bacteria were anaerobes, organisms living in the absence of oxygen. The bacteria catalyzed their activity by converting organic matter they fed on into storage molecules that decompose and release energy.

Artistic image of first bacterium

Circa four billion years ago

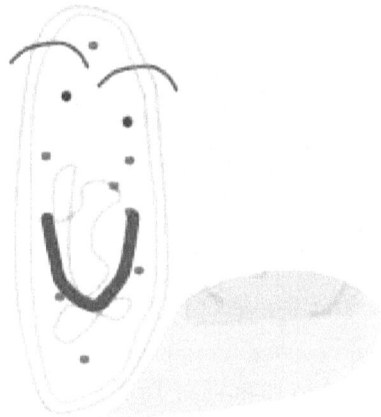

Reputed primary grandparent of all life forms

Some bacteria eventually became self–sustaining by synthesizing chlorophyll and using it with the sun's energy in the process of photosynthesis. The photosynthetic bacteria, mostly blue–green algae, combined earthly materials with water and carbon dioxide and released oxygen into the environs. Consumer bacteria ingested other bacteria, organic particles and waste products. People would not exist were it not for those early microbes living upon the juvenile earth.

Bacteria reproduce by binary fission, a simple cell division that is asexual, or without the sexual union of male and female gametes such as sperms and ova. When a bacterium reproduces, it duplicates its single chromosome and divides into twin daughter cells. Each daughter cell receives a replicated chromosome, which makes them genetically identical clones. Binary fission doesn't provide genetic variation, as does the sexual union of gametes. Bacteria obtain some genetic diversity through several other processes.

Bacteria didn't select sexual reproduction that combines different sequences of nucleic acids by two gametes merging. The tiny organisms are neither male nor female, and without gametes have remained sexless 'its'. Though bacteria can become genetically varied, the variation is limited without gametes. Still, bacteria obtain new traits by mutations that are changes in the nucleotides of genes. A mutation is an altered gene, or a new DNA sequence. In ten drops of bacteria there are one billion (1,000,000,000) individuals. Statistically, each drop has a potential of 35,000 mutations and viruses have a high rate of mutations.

Mutations may be caused by mutagens such as x-rays emanating from cosmic rays. The rays are high-energy charged particles entering the atmosphere from outer space. Cosmic rays penetrate the earth and return as x-rays, a type of secondary radiation. When x-rays pierce living cells they can alter a nucleotide sequence producing a mutation. Most mutations are unfavorable and are discarded by natural selection. Beneficial mutations are used by natural selection for fitness in promoting survival and reproduction. Neutral mutations are left alone or eliminated according to the degree of neutrality.

A bacterium facing bad conditions may mutate and start a new colony by rapidly dividing 1-2-4-8-16-32---and so on. The new colony might or might not be able to survive the bad conditions. Provided the new colony starts dying it might mutate into a colony that can tolerate or thrive and reproduce in the troubling environ. Survival and reproduction is the name of the game played by all living things. Pathogenic bacteria called 'bugs' have played many a winning game with rescuing mutations that produce strains able to withstand antibiotic drugs.

Bad Bugs

A hypothesized mutation rescuing some bacteria occurs in a pond where subterranean changes were increasing thermal activity. The temperature was rising and the bacteria needed an insulating cell wall. So far, their methods of obtaining genetic diversity had not bestowed the genes for the cell wall. The prospect of the half-baked bacteria escaping with their lives lay in 35,000 potential mutations and their rapid reproductive rates. The executioner's axe was about to fall on the bacteria gasping for diffusion. Suddenly, a reprieving x-ray emanating from cosmic radiation induced a mutation in a bacterium for a thicker cell wall.

The favored mutant reproduced until it multiplied into a colony of heat resistant bacteria. The sole escapee had been a clone, or an 'identical twin' in the perished colony. In the new colony it was a genetically altered clone the same as the other clones. In a sense, the old colony was not saved, but died and was replaced by bacteria with a novel trait. Had the temperature decreased the bacteria might have traded their thick cell wall for a thinner wall. The thin wall would return the colony to a semblance of its pristine genetic motif.

CHAPTER 3

Cloning works fine to propagate bacteria and boost their endurance, but multifaceted organisms depend on sexual reproduction to prevail generation after generation. Sexually reproducing organisms offer natural selection diverse choices of genetic traits to accept, reject or leave alone. A sixth finger on a person's hand, for example, is harmless and is left alone by natural selection. On the other hand, many organisms have been confronted by situations that were harmful.

One harmful situation, for example, proposed by some scientists was the declining birth rate that doomed the Homo species Neanderthals to extinction thirty thousand years ago. Another proposed malefactor was a lack of new mutations to increase their genetic diversity and give Neanderthals a better chance at redemption. The problem with waiting on new mutations in desperate situations is the mutations usually need to produce a severe change to do much good. Organisms resist severe changes because natural selection has stabilized their genotypes and they are 'settled in their ways'.

There are other plausible proposals for the demise of Neanderthals and one is contemporary Homo sapiens were superior hunters and took the lion's share of animals. The reduction of animals deprived Neanderthal tribes of essential dietary proteins, which led to susceptibility to diseases and declining reproduction. Judging by the brutal history of human warfare, it's easy to imagine Homo sapiens displacing Neanderthals in conflicts over resources. However, some scientists tell the puzzling story that people started wars after Neanderthals became extinct. A cheery theory is

Neanderthals didn't vanish, but had their pleasures by
peacefully interbreeding with Homo sapiens.

Imaginative portrayal of a Homo sapiens

Romancing a Neanderthal

A cheery story and happy ending for some threatened
Precambrian bacteria is revealed in fossil evidence indicating
they networked in an ocean region dwindling in nutriments.
Famine followed and endangered bacteria formed colonies
that acquired flagella, tiny whip like structures. The colonies
foraged for nourishments by paddling around on the tiny
appendages, plausible precursors of pili that are short
tendrils on extant bacteria. Flagella are assumed adaptations
by Precambrian bacteria to cling to grains of sand. The wee
beasties plodded along the old ocean bottom on the tiny
limbs like creeping caterpillars.

The gene or genes producing the teensy attachments in a
famished bacterium might have mutated and brought about
longer flagella. That being the case, the enfeebled colonies
were salvaged by natural selection favoring the one
flagellating bacterium that reproduced profusely. Evading
starvation, colonies of foraging bacteria transmitted data to

other colonies by chemical effusions, gene fragments and equivalent devices.

All the foraging colonies that found nourishment broadcasted their success to needy colonies that followed in their wake. Unsuccessful foragers that were perishing dispatched signals to 'unwary' colonies to avoid the barren area and spared them with their genes intact. A cheery story and happy ending no doubt, but meaningless to organisms that act on instinct and are not conscious of their existence. Indeed, it would be a long time before another organism had the intellectual capacity and sensitivity to write, "She walks in beauty, like the night of a thousand cloudless climes."

Mindless organisms surviving bad times are sometimes portrayed as socially intelligent and strategists, but those qualities usually refer to cerebral animals like elephants, chimpanzees and humans. Thought processes of organisms whose brains have small cognitive areas are hard to evaluate. They would be easier to evaluate with better descriptive adjectives and phrases in the lexicon of science. Assessing how much animals less cognitive than people appreciate beauty or lasting affection for sexual partners is improbable. Indeed, when an elephant looks at a beautiful sunset or a potential copulatory companion it can't be said if the elephant admires either one.

Mindless organism

Duh?

Two Cognitive Mammals

Homo sapiens	Loxodonta africana
The English Bard	The African Elephant

Shall I compare thee to a summer's day?	I really can't say

Uninformed people living when scientific data was scarce couldn't view natural phenomena objectively, which is somewhat true today. They viewed science with superstition and fear like ignorant people view things that are different or events they don't understand. An ignorant crowd ridiculed Quasimodo, the hunchback of Notre Dame. Galileo was another victim of ignorance and barely missed a hot foot for insinuating the earth revolved around the sun. The frightened theocrats put Galileo under house arrest for the rest of his life. Ignorance and its offspring, superstition, are stumbling blocks that interfere with conceptualizing theories of evolution.

The adjectives creative and ingenious are anthropocentric ornaments that attribute human traits such as speech and behavior to nonhuman organisms. As 'gifted' and 'clever' as bacteria are, they respond without comprehension and would score a zero on an IQ test if they could take one. A communiqué from a failed colony that saves another colony is 'tactical' and scientifically defined as strategy, but their 'smart' move is a genetic directive. Anthropomorphizing chimpanzees is another matter since their behavior is so much like human behavior.

Conceivably, the genetically directed bacteria were aided by another innate mechanism that is activated by impending doom. A hypothesis poses the genotypes of living things contain gene sequences that induce an auspicious mutation in troubled times. The existence of an obliging mechanism would have furnished longer flagella for the hungry colonies. Assuming organisms have a genetic guardian, it is unreliable since it doesn't always come to the rescue. Mayr says, "needed mutations may have failed to appear when there was either a change in climate or the sudden arrival of a new competitor, predator or pathogen."

The failure of a useful mutation to surface when needed is the reason random chance, and not a pre-installed gene set is accepted as sparing or neglecting jeopardized organisms. The odds were good that the starving bacteria wouldn't have to wait long for a helpful mutation. Their vast numbers and quick-change artistry empowered the afflicted colonies to be redeemed by new nucleotides. Bacteria have lived long and prospered in contrast to 99.99% of all species that ever lived which natural selection or any evolutionary mode didn't spare from extinction.

The simplistic and swift reproductive capacity of bacteria made the first full-fledged microbes of the Precambrian the most enduring organisms of all time. Evolutionary biologist Stephen Gould says bacteria have existed for "two-thirds to five-sixths of the history of life." Measuring the success of the brainless bacteria, it seems nature treated them well. Still, good or bad treatment for any organism isn't etched in fossils since the stories they tell are usually incomplete.

Certainly, innumerable bacterial strains will never be known because their fossils haven't been discovered, the reason some extinct hominids aren't known. Perhaps any unknown hominids vanished because Nature didn't help them like she helped the starving bacteria, and if she did help, it wasn't good enough. Perhaps a contagious disease struck the unknown hominids and they didn't have the constitution needed to survive. Maybe some apes that left the trees and became bipedal hominids didn't adapt suitably to succeed. Many causes could have taken various hominids out of the picture leaving their fossils undisturbed in their earthly tomb.

The well-known hominid Homo erectus left plenty of fossils, but they don't tell why Erectus faded out. One hypothesis is their birth rate fell below their death rate so long they died out. Conversely, they possibly evolved into another species by pseudo-extinction; a theory posing many extinct species didn't vanish, but evolved into another species. Provided Homo erectus 'pseudo-extincted' means Erectus didn't fade away, but evolved into Homo sapiens. An analogy is the fictional Dr. David Banner turning into the Incredible Hulk. Thus, David didn't fade out, but evolved into the Hulk by special effecting pseudo-extinction.

Evolution of Homo erectus 1.8 to .5 million years ago

Brain expanded in spurts over 1.3 million years

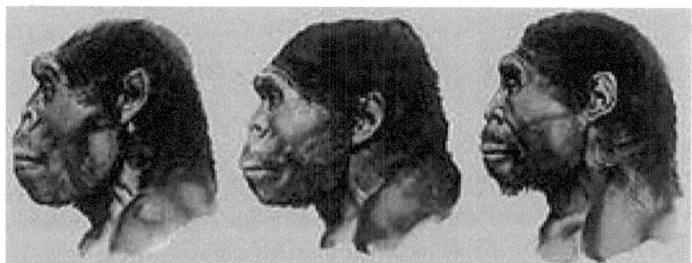

Getting smarter all the time

Still, if Erectus wasn't soaked up by pseudo-extinction suggests Homo ergaster gave rise to Homo sapiens. This being so, in time Homo sapiens displaced its contemporary

Homo erectus and the remaining Erectus precursor species, Homo ergaster, if any existed. Untangling the lines of descent, but hardly proving anything, scientists might one day agree that Erectus and Ergaster are the same species. Classifying and concluding what happened to extinct hominids is debatable and subject to change. What is certain is Home sapiens is a valid species that is rapidly reproducing and proliferating.

A characteristic of Homo sapiens is they inherently value the qualities of love, compassion, kindness and charity. People show the qualities in altruistic, or unselfish deeds. Practiced consciously by large brain organisms and instinctively by brainless organisms, altruism is a romanticized support mechanism that involves a slight to sizable cost to the altruist. The English writer, Charles Dickens, aggrandizes human altruism in *A Tale of Two Cities*, telling of a man who takes another man's place on the guillotine. "Greater love hath no man than this, that a man lay down his life for his friends." (John 15:13)

Far less costly deeds are chimps sharing fruits or people sharing jellybeans and salespersons handing out gifts to entice customers. Altruism was rooted in the genotypes of the early bacteria and is entwined around the bush of the animal kingdom. The extensive practice of altruism among organisms implies that it is favored by natural selection and genetically guided.

Evolutionary biologist Richard Dawkins theorizes altruism is biologically determined to preserve genes. He says the bodies that house the genes are throwaway survival machines. Dawkins proposes selfish genes discriminate and exploit their expendable bodies to save like genes, albeit their project could require the ultimate sacrifice. People and chimps have been injured or died while saving their relatives or other unrelated members of their species. Soldiers have been killed throwing themselves on hand grenades to protect their imperiled comrades. Stories abound about people making sacrifices for their family and friends.

Goodall tells of a chimp jumping out of a tree when approached by hunters, but returned to save her child and was shot dead. "Greater love hath no mother ape than this, that she would give her life for her son." (Sacred Scrolls 15:13) Since her son shared half of her genes, her sacrificial act served to protect her motherly investment. Despite the cost, when her son reproduces half of her genes will be dispersed, and a fourth more through each of his offspring. Credibly, altruism is biologically determined and interwoven throughout the biological kingdoms from bacteria to Homo sapiens.

Selfish genes inherited from the Precambrian bacteria presumably guide altruism in the biological kingdoms. Better to give than to receive is a verbal expression of an instinct several billion years old. The rationale for the old instinct comes from molecular studies that revealed the great age of many genes. People and bacteria share two hundred genes labeled 'old' genes. "Indeed," Mayr says, "it seems possible to trace some genes all the way from animals or plants to bacteria." Organisms using twenty essential amino acids to construct larger protein molecules exemplify related genetic processes.

Provided certain biological processes, such as altruistic behaviors are directed by similar or old genes, it presupposes what benefited the early bacteria is good for all living creatures. Still, a deed like jumping into a river to save a stranger is counterproductive because the Good Samaritan might drown. Dying and losing a throwaway body before its genes are dispersed isn't in tune with the theory of the selfish gene. Perhaps the reason for the ultimate sacrifice is that anyone in distress spurs a primeval instinct in people to safeguard the tribe.

Genetically guided deeds that safeguarded the tribe can be corrupted and detoured toward a 'scratch back' or fair-weather friend tactic. Back scratching is self-interested and is typical of someone who does another person a favor just to get a favor in return. Insincere back scratching is not the 'friend in need is a friend in deed' kindly act touted by idealists. Scratching backs is definitely in tune with salespersons giving gifts to soften potential customers. Clearly, political scratching isn't altruistic, but it is expedient and opportunistic. The biggest back scratchers in American politics are campaign contributors and lobbyists.

The opportunistic motive of back scratching by politicians is not why chimpanzees groomed each other in their natural habitats. Grooming in their undisturbed abodes was a no strings attached arrangement, rendered freely in the spirit of equality. Now, their habitats have been corrupted by human encroachment and the chimps, like people, are no longer in their adaptive social organization. Even so, grooming is still practiced in chimpanzee communities and is a soothing diversion that may go on for hours. Grooming removes insects, potential sources of disease, which suggests it is adaptive and genetically influenced.

Grooming is practiced for the same reason by captive chimps, and sometimes to gain favors. Captive chimps in a nonadaptive social setting, like people, form hierarchies of dominant and subordinate individuals. At times, subordinates grooming superiors, is a maneuver to procure sex. Captive chimps are restricted in movement, which isn't a principle adapted for the original chimpanzee social structure. Instead, chimpanzees adapted fusion and fission meaning to freely join or leave a group. Curbing their choice to come and go has a corrupting effect on the adaptive purpose of fusion and fission.

Grooming in Captivity

Pleasure or Politics?

CHAPTER 4

While the primary ancestors of chimpanzees effectively survived and reproduced, they were the only tenants of the Precambrian seas with every life process. They were as snug as bugs in rugs and during their lengthy span of snugness, or stasis, natural selection initiated few changes. The stasis signifies the temperature, salinity, and gasses diffusing between the water and atmosphere were essentially unvaried. The stability of the old oceans ruled out a reason to change, but if their stases had been interrupted, the bacteria had the potential mutations to change quickly.

Sustained stases, with spans of little change in life's history are commonplace. "The stasis of nonchange of most fossil species during their lengthy geological life spans," Gould says, "had been tacitly acknowledged by all paleontologists, but almost never studied explicitly because prevailing theory treated stasis as interesting nonevidence for nonevolution.

Evolution had been defined as gradual transformation in extended fossil sequences, and the overwhelming prevalence of stasis became an embarrassing feature of the fossil record, best ignored as a manifestation of nothing (that is nonevolution). How can we claim to understand evolution if we only study the percent or two of phenomena that constructs life's directional history, and leave the vast field of straight-growing bushes – – the story of most lineages most of the time – in a limbo of conceptual oblivion?"

A long period of 'nonevolution' was revealed in reef-building corals that changed little in 80 million years during the Cretaceous Period. Gould says there was "little evidence of directional change, but rather a story of oscillation within a range set by minimal and maximal size of corallites (individual coral animals within the colony)." On one side of their little evolving bush were larger corralites in turbid water that ate tiny animals. On the other side were smaller corallites living in clearer water around the reef top that fed on tiny plant algae.

The corallite animals were maintained at optimum sizes by natural selection removing corallites that grew too large or too small. The same stable environment nourished the two niches and selection picked the average traits within each group. The corralites remained in a period of stasis by gently seesawing between selection's size limits without a change in direction. Clearly, in the style of the Precambrian bacteria, the corralites, like many other static organisms were snug like bugs in rugs for a long, long time.

Colony of corallites.

Prolonged snugness

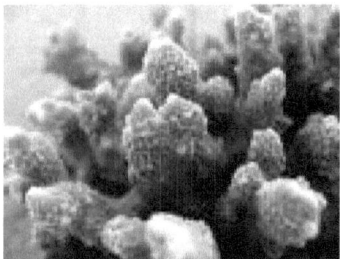

Ho Hum

Surely, there's no harm in static creatures loafing around epoch after epoch unless something brainy has a master plan for progress. People's numerous ancestors throughout geologic history couldn't have existed had they lived the ho-hum life of the corralites. Something had to happen to goose all those ancestors such as worms, fishes, reptiles and apes on their unintentional journeys. There's a rub in a sluggish species giving rise to a new species. A sleepy species has to be rousted like rousting Rip Van Winkle out of bed with a firecracker. Mostly it's the environment that causes evolutionary rousting.

The weather, new predators, pathogens, mutations plus other environmental factors initiate and set the pace of modification in organisms. A rapid climate change, for example, may cause a species to adapt hurriedly to the surroundings. Adapting can produce a new species or simply modify a species responding to an altering climate or to another evolutionary mode. Moreover, abrupt changes in a static species followed by an improved species are noticeable in the fossil record.

The rapid replacement of species in stasis was addressed in 1972 by paleontologists Gould and Eldredge in their Theory of Punctuated Equilibrium. A static species is inactive and not evolving, and is said to be in equilibrium. Replacing the species rapidly gives the concept of Punctuated Equilibrium its name. Opposing the theory, some scientists think abrupt discontinuity and speedy speciation shown in the fossil record is incorrect because missing fossils cause gaps that obscure gradual changes.

Despite whatever causes the appearance of discontinuity and speciation in the fossil record, the pace is relative and measured in many thousands or millions of years. There is no standard to gauge speciation and evolutionary changes as fast or slow. A species that evolved from its primary ancestor in five million years would be rapid compared to a fifty million-year evolutionary journey.

Humans and whales are examples of 'fast' and 'slow' evolution. Homo sapiens evolved rapidly from the ancestral ape in five million years while whales evolved slowly over fifty million years from their primary ancestor. Some trees live long lives of thousands of years while people usually die in less than eighty challenging years, for what their brief lives are worth. "Out, out, brief candle! Life's but a walking shadow, a poor player that struts and frets an hour upon the stage and then is heard no more." (Shakespeare)

Also heard no more except in his writing is Ernst Mayr, but while he lived he explained discontinuity followed by the swift rise of improved species in his Theory of Specialization Evolution. Mayr says a population diverging from its parent population can become a new species by undergoing "a profound genetic restructuring." Profound is a relative term and wouldn't apply to genetic differences among people, which are superficial. A hypothetical instance of genetic restructuring is a few zebras separated from their parent population. A million years later the zebras were slightly different genetically, which slightly changed their phenotype.

What identifies a zebra population, or any animal population such as frogs or people as one species would change unless animals exchanged genes to stay in genetic equilibrium. Obviously, the early hominids that migrated to northern Europe and adapted white skin didn't at that time exchange genes with those in Asia that had adapted bronze skin and Asian eyes. The degree of genetic restructuring determines the degree of change in a species, and people have very little genetic restructuring. Mayr's Theory of Specialization Evolution ties in with the Theory of Punctuated Equilibrium from a genetic perspective.

Asian eyes

Ecological adaptations

Oriental Homo sapiens

Tan Skin and Asian Eyes

Genetic restructuring by punctuated equilibrium?

Punctuated equilibrium or Speciational Evolution might account for the rapid rise of Australopiths from arboreal apes had the apes been in stasis. However, rapidly turning ancestral apes into chimpanzees and later into Homo hominids by profoundly restructuring their genes is confusing. The uncanny similarity of human and chimpanzee genotypes baffles the scientific community, which offers disparate theories. The similarity of the two primates implies Speciational Evolution doesn't account for the rapid evolution of the Homo lineage ending with humans. What on earth initiated Homo sapiens speciation is hypothetical, assuming it was on earth and for what it's worth.

Undeniably, the sudden appearance of the first hominid fossils means something or other punctuated those apes right out of the trees. Whatever it was left most of their genes alone, ironically turning the apes swiftly through the Homo line into the peculiar human species. Arguably an event if left undone would have kept many suffering souls in conceptual oblivion, whatever constitutes a soul. Nevertheless, a billion years ago events that were not left undone kept people out of oblivion and into temporary consciousness.

Events that occurred in the Precambrian seas long before people saw the light of day gave birth to the protists, second ancestor of the succeeding biological kingdoms. The characteristics of the first protists, like the characteristics of the first bacteria, are inferred from their remnants and existing species. Protists are larger and have more parts than viruses and bacteria. Microbiologist Lynn Margulis believes protists evolved from colonies of bacteria. Conceivably, the members of some colonies of foraging bacteria combined into larger bodies to facilitate absorption of nutrients and the bodies became protists.

Another supposition poses protists arose from bacteria that were penetrated by x-rays causing them to mutate into larger bacteria. Luckily, the larger bacteria filled the need for a different surface to volume ratio allowing them to acclimate to varying concentrations of atmospheric gases. Natural selection favored the larger mutants, but details of the big bacteria evolving into protists are speculative, as is the speciation of any organism. Theories explaining modes of evolution and speciation are formulated from extensive scientific data. Testing the theories gives their conclusions the highest probability to be correct.

One specific occurrence considered essential to evolution continuing after the advent of bacteria is explained by the ecological theory of the cropping principle. Accordingly, bacterial algae monopolized the Precambrian waters and other organisms 'trying' to evolve, couldn't get a foothold. These environs "have an enormous biomass," Gould says, "but they are usually impoverished in numbers of species." Fortunately for optimistic people happy they were born, a species of herbivorous protists called croppers arose in the ancient seas.

A new species like the croppers is instinctively driven (biologically determined) to increase its survival and reproductive potential by exploiting untapped regions. The impoverished biomass was untapped, and applying a microscopic version of Manifest Destiny, the cropping herbivores starting eating the algae like rabbits in a cabbage patch. The microbes feasted on "the abundant species," Gould says, "thus limiting their ability to dominate and freeing space for other organisms. Croppers made space for a greater diversity of producers, and the increased diversity permitted the evolution of more specialized croppers."

A hypothesis poses the first croppers arose from a mutant strain of bacteria that evolved into protist herbivores. The bacterial strain could have emerged from a bacterium whose genetic material was zapped by x-rays originating from cosmic rays. A galactic baptism of the croppers portends a celestial beam christened the Crown of Creation. Let the Crown have dominion "over every creeping thing that creepeth upon the earth." (Genesis 1:26) Unfortunately, the things that creepeth don't have much of a chance against the Crown with the big brain.

Whether earthly things creep or leap it helps to conceptualize their intermediate or transitional stages since an existing species can also be labeled a dead-end. Homo erectus, for example, was a dead-end if it did not give rise to Homo sapiens. However, if Erectus did spawn Homo sapiens, Erectus would be an intermediate or transitional hominid between Homo habilis and Homo sapiens.

The series Habilis-Erectus-Sapiens or A-B-C is an evolutionary progression accepted by many, but not all scientists. People as C are intermediate animals if they generate a new species. C-X represents humans producing a new species, X, with unknown characteristics. Presumably, the X species would approximate its precursor, perhaps completely hairless, a bigger brain, teensy testicles and mammoth mammary glands. People are dead-ends if they stop evolving or become extinct, credibly a blessing for themselves and clearly better for many vanishing species. The theory of natural selection holds that whatever comes to pass will be random, which applies to a mass extinction.

Homo habilis posing with a stone tool.
Brain about 20% larger than a chimpanzee brain

Say cheese

Another random production of Nature is the otter that looks like an intermediate 'half' or 'fourth' animal. Otters are adapted to water and land and might be at any evolutionary stage such as 1/4, 1/2 or 1/8. Whatever stage they're in, they're dead-ends if they do not produce a new species. An evolutionary stage isn't a predestined progression unless there is a mystical motivator, which scientific consensus rejects. Provided otters are still changing, events are too slow to observe, but if otters continue adapting to water they might become as aquatic as whales, complete with a blowhole and flippers.

Otters, dolphins and cows are mammals that diverged from the common ancestor they share with people. The three mammals are physically adapted to three different environments. Cows are the female gender of cattle that prehistoric people artificially selected from an obscure ungulate. Milk cows are domesticated animals that were bred until their udders, or mammary glands became enormous. Dolphins are marine animals and both are people favorites such as Flipper and Daisy. Mother Nature would be challenged to adapt otters, cows, dolphins and humans for flight the way she made some dinosaurs into flying reptiles.

Two Affectionate Otters

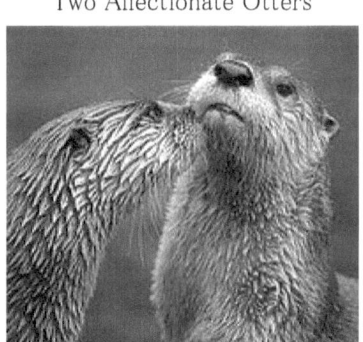

A Dolphin and Cow in a Nonadaptive Environment

The common ancestor of otters, dolphins, cows and people was among countless new species, intermediates and dead-ends following the Croppers Feast. Physical and biological changes caused multicellular animals to proliferate during the Cambrian Explosion, a geologic moment of a few million years in the Cambrian Period. The biological changes are attributed to cropping protists removing excessive algae that freed space for other species. When the explosion ceased, nearly all categories of animals had evolved. Since then, their basic constructs were modified, but with few new designs. Gould says, evolution "only recycled the basic products of its own explosive phase."

A physical explanation for the explosion is it was initiated by an increase in the amount of oxygen required by complex organisms. The simpler Precambrian organisms used oxygen for respiration and ozone to absorb harmful ultraviolet rays. The oxygen was slowly released by multitudes of blue-green photosynthetic algae. Eventually, the oxygen level necessary to sustain multicellular animals was reached. The breath of fresh air triggered the explosion and the animals rapidly evolved to radiate into vacant niches. Some researchers disagree with the explanation saying the algae had supplied ample oxygen for multicullular animals before the explosion.

Many of the early multicellular animals could reproduce asexually, and by sexual reproduction that was a pivotal protist 'invention' that increases genetic diversity. Sexual reproduction is so efficient that the likelihood of two people being genetically the same is one in seventy trillion. Increasing the number of diverse genes enables more adaptive changes and renders new species and subspecies. Uniting sperms and eggs to produce offspring is an adaptive strategy of sexual species to enhance survival and reproduction.

Conversely, asexual reproduction occurs without gametes and is great for bacteria and potatoes. However, asexual cloning limits the complexity of organisms Nature can construct. Perhaps asexual reproduction was once satisfactory when a need arose for sexual reproduction. It's also possible sexual reproduction wasn't needed, but began by a spontaneous mutation. So far, asexual reproduction hasn't produced cognitive organisms except for science fiction. "The Thing," for instance, was a wicked seven-foot vegetable man who landed on earth to cause trouble, but the good guys fried him to a crispy critter. Apparently, he came from a planet where evolution took another twist.

On planet earth, individuals with the best genotypes have the greater chance to survive and reproduce more offspring. Plenteous progeny gives their species better survival odds. Many people in modern society choose to prevent or limit offspring by separating sex from reproduction, a practice simplified by efficient contraceptive methods. Fruitless wombs keep would-be children in conceptual oblivion, which some pragmatists advise is the best option. Inside that conceptual nothingness are no troubles or fear of a dashed hope of life everlasting. Permanently ending human reproduction is a noble alternative, especially for the rest of the organic world.

Theory holds simple sexual reproduction was invented by the first protists to increase individual differences and thus abet their survival. Protists also invented the nucleus that is a little body within the cell's enclosing membrane. Nuclei contain chromosomes and other genetic information necessary to control cell growth and reproduction. Although scientific consensus maintains evolution is not designed, the nucleated protists were crucial to later organisms. Any organism following the inventive protists would be genetically restricted without sexual reproduction and up a stagnant creek without a nucleus.

Special structures that supposedly evolved by symbiosis in the protists are called organelles. The structures are assumed precursors of organs, implying protists jumpstarted the long formative journey of hearts, livers, brains and other organs. As 'advanced' and essential protists were to future evolution, they lacked the specialized cells and tissues that were added piecemeal to later organisms. Despite their 'shortcomings', the protists played a great survival and reproductive game side by side with the viruses and bacteria. Presently, protists are represented by more than 8,000 descendent species that left most past life forms in the Hall of Extinction

Diatom

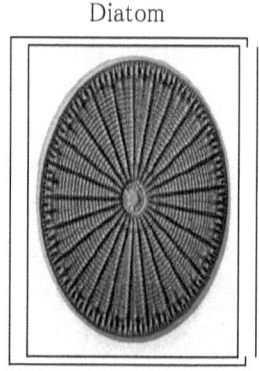

Diatoms are unicellular protists, but may occur as colonies. They usually reproduce by cell division, but sometimes by sexual reproduction. Many are vital to marine life as components of plankton in the oceans. They are distant cousins of people.

Some protists are flagellates that move by flagella, conceivably produced by genes akin to 'older' bacterial genes that fashion pili. These 'old' genes could be coding genes for the tail on human sperm cells. Adding parts while progressing from small to large seems like a contractor's plan for a house. However, some blueprints pattern creatures that look like the bathroom was built on the roof. The anus of a starfish, for instance, sits on top and its mouth is flush on the bottom. Really topsy-turvy is a jellyfish that doesn't have an anus and uses its mouth to dump wastes.

Protists added other parts and devised the mitochondrion, a multipurpose little structure that is a component of cells with a nucleus. A mitochondrion is located outside the nucleus and contains its own DNA. Mitochondria have some features resembling those of bacteria. Many scientists hypothesize mitochondria evolved from small bacteria that were engulfed by larger bacteria. The two bacteria formed a symbiotic relationship, the cooperative practice of many organisms and it time the small bacteria evolved by symbiosis into mitochondria.

There are several kinds of symbiosis; commensalism is an association between two organisms when one benefits and the other is unaffected. Mutualism is an association between two organisms where both benefit. Algae and fungi, for example, survive extreme conditions by living together as lichens. Parasitism is an association in which one organism benefits to the detriment of the other. A tapeworm is a parasite that feeds on the digesting food of an animal.

"In the discussion of evolution," Mayr says, "not nearly enough attention is paid to the overwhelming role of symbioses." Provided a symbiotic relationship gave rise to mitochondria, it and other cellular components of protists were passed down through the biological kingdoms like cherished heirlooms. The advent of the mitochondria moved the ingredients of the future human genotype out of the evolutionary stage of the Precambrian bacteria. Lots and lots of advents would be needed to build a human body.

Mitochondrion

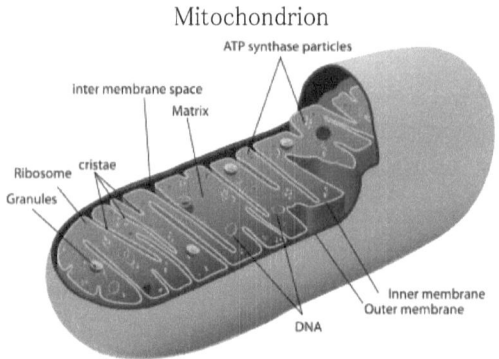

The mitochondrion was small, but an essential part of the first protists and vital to the evolution of subsequent organisms. The tiny organelle has a star role in the Out of Africa or Mitochondrial Eve Theory of human origin. The male and female mitochondrial DNA of humans and other sexually reproducing organisms is located in the egg and tail of the sperm respectively. The tail doesn't enter the egg at fertilization so the baby only inherits the mother's mitochondrion leaving the tail with the male's mitochondrial DNA out in the cold.

Geneticists compared samples of human mitochondrial DNA taken from various ethnic groups all over the world. They theorized from the findings that Homo sapiens originated from an unknown hominid in Africa 140,000 to 290,000 years ago. The higher figure agrees with some researchers who think early Homo sapiens speciated 300,000 years ago, for what it's worth. Scientists traced the mitochondrial DNA to the first Homo sapiens ancestor who was a woman they named Mitochrondrial Eve. However, they realized other women were alive at the time and reasoned their lines of maternal inheritance had died out.

Tracing samples of mitochondrial DNA enabled the scientists to construct a family tree. The African branch of the tree is the oldest and longest and it sprouted all the peoples of the earth. The tree shows the time periods people started becoming slightly dissimilar in their genetic composition. As African populations began to migrate and radiate in stages, they entered and exploited new environs. Consensus is their migratory paths followed food resources such as animal herds. They hadn't gotten into colonizing and dominating people since there were no people along the paths to colonize and dominate.

Responding to environmental variations where they settled, the African emigrants became outwardly different to a minor degree. Their physical or phenotypic distinctions were superficial traits, most notably in pigmentation. Two hundred thousand years later, give or take, they developed diverse languages and cultures and an unplanned economic system driven by direct competition. The acquisitions plus their physical differences and atypical huge brains pummeled the poor things with "the thousand natural shocks that flesh is heir to." Still under a constant pummeling, people keep multiplying with no possibility of individual survival.

World War I soldier pummeled by an
unnatural shock of mustard gas

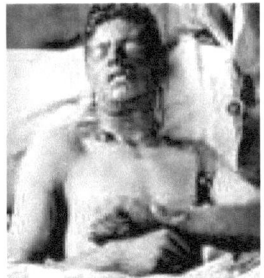

So much for good will to men.

CHAPTER 5

The thousand natural shocks couldn't pummel people until the poor things evolved. What's more, it took lots of evolutionary events to get them to the pummeling block. A crucial event that occurred before the earth formed was a supernova exploding and spewing gasses that contained people's chemical makings. The gasses drifted through space and ultimately condensed into the solar system. When people's particles got into the Precambrian seas, they still had to pass though bacteria and protists to finally jell into humans. After passing through protists their voyage through geologic time became difficult to charter in the fossil record.

Meanwhile, more voyages of Precambrian organisms such as plants and funguses were launched by the innovative protists. Plants reproduce asexually or sexually according to the species. They are theorized to be produce that grew out of a symbiotic relationship between protists and photosynthetic bacteria. Their supposed descendants, the funguses or fungi, are parasites that do not make their own nutrients by photosynthesis. Ringworm that causes Athlete's foot and mushrooms are two familiar funguses and distant human relatives.

Mushroom Cousin:Portobello

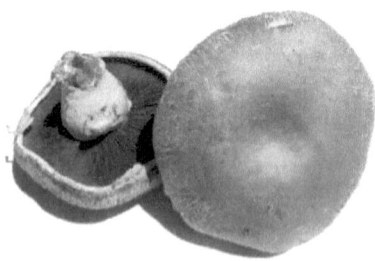

One hypothesis for the origin of funguses is some plants drifted into deeper ocean waters with little sunlight. Later, two plants formed a mutual symbiosis and one became dependent on the other. The dependent plant lost its ability to photosynthesize, and during millions of years evolved into funguses. This being so, illustrates the important role of symbiosis in evolution.

Funguses and many bacteria are decomposers that 'eat' energy enriched organic molecules previously synthesized by producer plants. The funguses and bacteria secrete enzymes on the plant molecules to reduce them to a digestible meal. "The major cycle of life runs between production and reduction," Gould says. "The world could get along very well without its consumers." Funguses might have become some sort of animal indicated by molecular analysis. The analysis showed their cell walls are composed of chitin, an insect component absent in plants. Mayr says, "In much of their basic chemistry fungi are quite closely related to the Animalia."

Funguses missing the target with the shot they had at being animals is one of the countless 'what if' incidents in evolution. The incidents highlight the role random chance plays with natural selection in determining if a species will come into being. A big IF that didn't happen is canceling the voyage of human evolution, or at least the big brain. That cancellation would be a blessing for the species suffering under human exploitation, and a relief for the suffering people exploiting each other. What ifs make trivial chitchat, but that's all.

Arbitrary what ifs did not cancel the Precambrian voyages that sailed on and seeded the extensive Tree of Life. A theory holds the tree's animal bush stemmed from colonies of consumer protists, called choanoflagellates. One indicator is, both choanoflagellates and the human animal use similar proteins in communication. Similar substances that operate in different organisms lend more credence that organisms inherit 'old' genes. The fact that people and choanoflagellates use the same or similar genes implies the early protists were human grandparents.

The Tree of Life

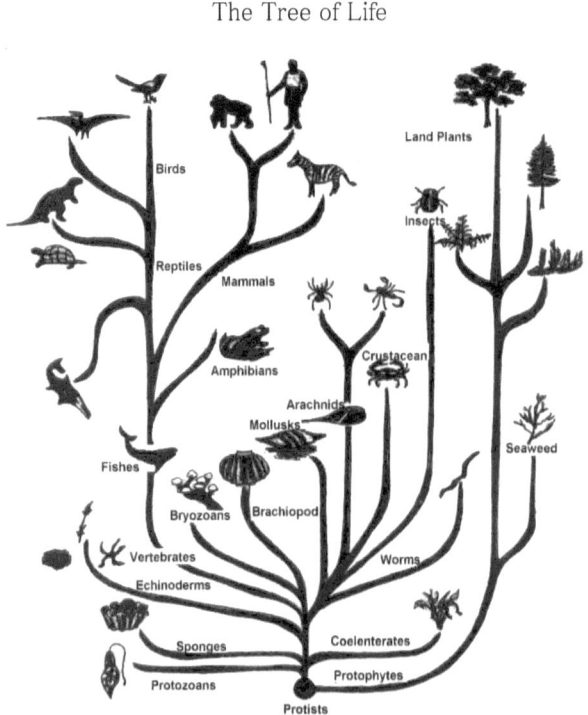

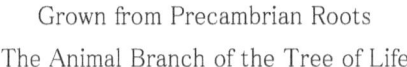

Grown from Precambrian Roots
The Animal Branch of the Tree of Life

Grown from Consumer Protists?

   Other evidence of genetic inheritance is choanoflagellates and sponges have common structures like collar cells inferring choanoflagellates gave rise to sponges. Choanoflagellate intermediate or offshoot species are not included in the series A–b–d–C, bacteria–protists–choanoflagellates–sponges. That is, organisms between d and C were intermediates if they gave rise to C. Organisms after C were offshoots unless they gave rise to another organism, which would make them intermediates. When species, intermediates and offshoots spread out as they evolve they form a bush.

Choanoflagellate showing collar cells

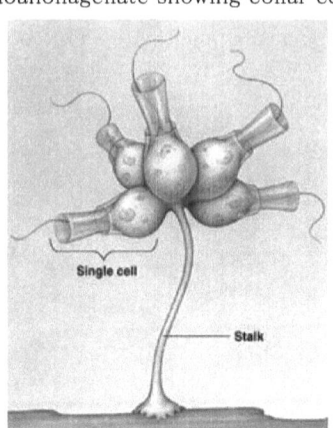

An early human seed?

The series A–b–c–D is not a time continuum because the offshoots of a common ancestor spread from each precursor in the series at different rates to form a bush or tree. Bushes and trees are terms often used interchangeably. The intermediates are stalks in the bushes or trunks in the trees, depictions best conceptualized as an overall picture. A series only represents organisms in the series and their direct precursor such as A–b–c–D, Homo habilis–Ergaster–Erectus–Sapiens. Homo neanderthalensis is the offshoot that supposedly diverged from Ergaster or Erectus, b or c.

The succession in the Homo line is unsettled, as are most lines. Some scientists question the validity of Habilis and Ergaster, and the place of other Homo hominids in the series. The Homo hominids are classed by their species name, when they existed in the fossil record and the average brain volume of a species in milliliters or ml (cubic centimeters). Homo sapiens are at the end of the series A–b–c–C analogous to sponges that are usually considered an end product. Should people and sponges meet extinction they would be dubbed dead-ends.

Therefore and presently, Homo sapiens and sponges are end products in the series that began with bacteria. Sponges as an end product denote they are human cousins and not grandparents. Still, there is a minor opinion that sponges are in a series leading to people, and if true, they are distant grandparents. Whatever sponges were, they were the first muticellular animals and, lacking a better adjective, the most primitive because they have the fewest systems to sustain the life processes. Nevertheless, they have sustained them a long time.

Sponges are invertebrates, animals without backbones. They have no digestive tract, no circulatory system, no muscle or nerve cells and labor is divided among cells. Plausibly, the single cells of their precursor differentiated into special cells that shared labor. They live fixed on the ocean floor and feed by absorbing suspended nutrients from water moved through their shapeless bodies by flagellating cells. Sponges are intuitively curious creatures and reminders that natural selection is aimless.

Sponges have Jacks-of-all-Trades amoeba-like cells that creep around the body exuding supportive substances for the skeleton. Most are hermaphrodites with both sexes that turn amoeba-like cells into sperms and ova. They cross-fertilize with other sponges and when their gametes unite they form an embryonic cell. The cell becomes a swimming larva that resembles a protist flagellate, credibly produced by old genes. Sponges also reproduce asexually from small internal buds. Sponges regenerate like plants or funguses, as do most primitive animals. When a sponge is torn apart each fragment grows into another sponge.

Although sponges are multicellular animals, some scientists see them as hardly more than a large colony of several different kinds of cells. They are one of nature's oddities by human perception and would have played a fantastic role in The Thing. Their few life support systems give the impression that a mission toward building a complete animal was interrupted. However, missions don't fit the theory of natural selection that holds evolutionary building events are random.

What accounts for the incomplete appearance of sponges is they've changed little since their evolution. One idea for the small change is natural selection was seldom pressured to act. Sponges live snugly on the bottom with currents catering to their nourishment. Their environs and lifestyle helped them hang around for more than six hundred million years and survive all catastrophic extinctions.

There are several thousand species of sponges in the large category of a phylum. Although they've changed little, they could hardly be called static or in equilibrium. They are tagged primitive by the Crown of Creation because they operate on few biological systems. Attila the Hun was primitive, but he outsmarted the patricians who governed the civilization of the Roman Empire. One 'smart' move made by some sponges that lived in nutrient depleted waters was to become carnivorous. They 'learned' to capture tiny crustaecans by entangling them with small threads and digest them with more enveloping threads.

Other 'savvy' sponges host photosynthesizing organisms as symbionts such as green algae that furnish the sponge eighty percent of its energy. That clever sponge built part of its skeleton to shine, so it could conduct sunlight for the photosynthisizing symbionts. These inventive and enterprising sponges make do with available resources like a hermit crab makes do with a discarded tin can or old shoe for its home sweet home. Despite all their entrapnureal accumen, not one sponge has sense enough to get out of the rain.

Clearly, there is nothing like natural selection and instinct to save the day.

Fossil sponge

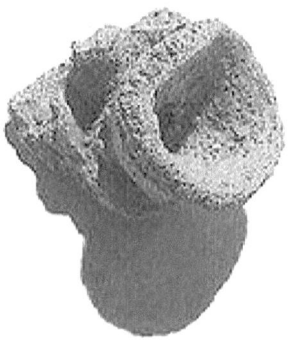

Resembles modern sponge

A marine mate of the late Precambrian sponges is the jellyfish that has a few more life support systems. The jellyfish is another weirdo that doesn't know enough to get out of the rain, but lives great by its innate 'wit'. The jellyfish phylum comprises nine thousand species and includes corals, the free-floating medusas with tentacles, and the Portuguese man-of-war. The man-of-war has dandling tentacles that can be as long as fifty feet.

Jellyfish usually reproduce by alternation of generations that occurs by sexual and asexual phases 'taking turns'. In the jellyfish life cycle the sexual medusa alternates with the fixed asexual form called a polyp. This kind of reproduction is limited to some plants, algae and a few invertebrate animals. At one time this reproductive strategy was the best, or perhaps the only method natural selection had to offer. Then again, like the protists, jellyfish might have chosen sexual reproduction to supplement asexual reproduction to increase genetic diversity.

While jellyfish and sponges were evolving, Nature was growing the vast bush of living things. However, there is no consensus in determining the evolutionary paths of the organisms in the ever spreading bush. A theory holds that jellyfish descended directly from choanoflagellates, its common ancestor with the sponge. Another holds jellyfish evolved from sponges or an intermediate animal derived from choanoflagellates that gave rise to jellyfish and sponges. Whatever theory is correct, choanoflagellates were workhorses in the Precambrian seas.

Regardless of their origin, jellyfish and sponges changed little in six hundred million years, and their stability is accredited to normalizing or stabilizing selection. Normalizing-stabilizing selection secured the corralites for a long time and it operates in relatively stable environments when traits rarely need to change. Apparently, sponges and jellyfish are in stasis because normalizing-stabilizing selection continually hones their genotypes to the environment. When there is no reason to change, natural selection lets sleeping dogs lie. The brainless sponge and jellyfish evolved low on the evolutionary totem pole at the start of the Cambrian Period.

Jellyfish Life Cycle

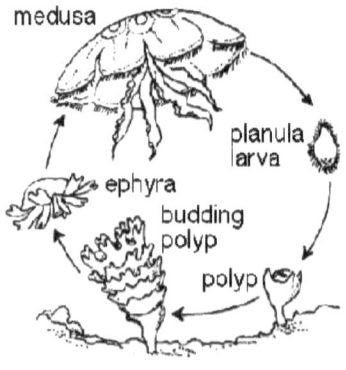

Alternation of Generations

Six hundred million years ago in the Cambrian Period the animal bush was firmly rooted. The waters of the earth brimmed with invertebrates and their relatives. The flatworm is a model for an ancient multicellular grandparent that evolved and diverged into thriving new species and varieties with life systems of sterling capability. However, the burgeoning progeny came with no warranty, and like other living things, most of the little animals fell to extinction. Many flatworm descendants became parasites, perhaps avoiding extinction by supping on the substance of helpless hosts.

Flatworm

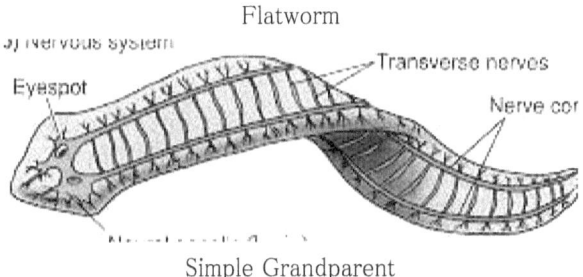

Simple Grandparent

The flatworm produced descendants with pulsating vessels and blood diffusing through open circulatory networks. There were worms with a complete mouth to anus digestive tract and a one-chamber heart pumping blood through a closed circulatory system. Coding genes steered by regulatory genes guided the embryonic growth and development of efficient nervous, respiratory, excretory, reproductive and muscular systems of the animals.

A notable genetic fact about worms, humans and other living things is that different cells perform different tasks, but they have the same kinds of chromosomes and genes. The profound differences between the cells such as muscle and nerves, Gould says, "does not arise from genetic constitution, but from alternate paths of development. During development different genes must be turned on and off at different times in order to achieve such disparate results from the same genetic system. In fact, the whole mysterious process of embryology must be regulated by exquisite timing in the action of genes.

To differentiate a hand from a homogeneous limb bud, for example, cells must proliferate in some areas (destined to be fingers) and die in others (the spaces between them)." The genes that turn other genes on and off and speed them up or slow them down are regulatory genes. Mutations of regulatory genes may cause drastic results in the offspring. The peculiar human body, for example, is believed to be a drastic result of a mutation in the regulatory genes. Apparently, minor differences in human and chimpanzee regulatory genes account for the major difference in their phenotypes.

Primitologist Jane Goodall and a chimpanzee friend

Two primates that should look alike

based on their nearly identical DNA

A mutation in a regulatory gene?

Human and chimpanzee grandparents bequeathed by the flatworm teemed in legions of invertebrates during the

Cambrian Period. These little animals without backbones proliferated throughout the old oceans and seeded one luxuriant bush after the other. The model for the diverging invertebrates is the roundworm that was more complex than its flatworm precursor. The anatomy of the roundworm was 'advanced' and 'better', implying a progression from an inferior to a superior status. Progressive terms stress the lack of adjectives to describe physical and chemical traits adapted from random genetic variation.

Then too, Mayr says, it "depends largely on what one considers to be progress." Planned progress is a student promoted to a higher academic level, but Homo erectus evolving from Homo habilis isn't considered planned or progressive. True, planned and progressive from simple to advanced organisms reduces such redundancies as the overused 'more complex', but the terms conflict with random chance. Regardless of which of the two worms was simple or advanced or more complex, these two common ancestors of all subsequent animals persisted and are represented today by numerous species in widespread habitats.

A simplistic hypothesis for the roundworm's speciation is a few flatworms became isolated in a slowly stagnating lagoon. The situation was dire because the water was slightly toxic and quick action was crucial. Natural selection rapidly favored traits best suited to adapt them to the unfavorable surroundings. Through countless years, intermediate worms emerged and one line gave rise to the roundworm. There's no telling what course evolution would take had the lagoon become intolerably toxic. However, if it did become intolerably toxic there might be fewer animals facing extinction.

Another simple hypothesis for the roundworm's speciation is some flatworms lived millions of years in a slowly changing lake. Their reproductive structures randomly prepared gene combinations that coded for traits with varying serviceability in the water. Individual differences offered natural selection diverse worms with their traits and coding genes. Some worms were more adaptable than others, and the best survived and were selected. Gradually, the worms produced a bush of intermediate worms and one branch grew into a bush of roundworms. One roundworm species became a human ancestor, an ill omen for multitudes of future flora and fauna.

The first hypothesis of worms in a stagnant lagoon accords with punctuated equilibrium by the rapid response of natural selection. In the second hypothesis natural selection is gradually favoring random traits and operating a step or so above normalizing selection. Gradualism is the mode of selection that slowly seesawed the corallites between optimum sizes. Fast and slow evolutionary events are relative terms that are exemplified by punctuated equilibrium and gradualism. Fast, for instance, is the micro-speciation or microevolution of fruit flies into a subspecies. Slow is the macroevolution of roundworms into amphibians or amphibians into mammals.

Roundworm

Flatworm's child

Two roundworm descendants are African and Asian elephants that are separated by multiple generations and untold years from their primeval precursor. The elephants are easily recognized as closely related species and appear products of gradualism. Yet, some elephants were hurried by necessity to adapt to an approaching ice age. Those with traits best suited for cold weather such as thicker hair and stockier bodies were chosen during successive generations. The elephants didn't have time to twiddle their trunks so selection had to act quickly else they chill out forever.

The hardier elephants evolved into mammoths whose fossils clearly show they were a type of elephant. The elephants were spared by natural selection 'turning' them into mammoths in a short time. Genetic variations, helped by useful mutations and random fortune gave them the right stuff to prevail while keeping that elephant look. In similar manner, two other descendants of the worms, the duckbill platypus and Homo sapiens might have emerged by accelerated selection using any trait with half a chance to work. Then too, it's credible these curiosities arrived by a rare and unspecified mode of evolution.

Additional progeny of the worms, the dinosaurs, were in the same boat as the threatened elephants and needed rescuing, but there was no redemption. Whatever caused the extinction of the big reptiles sixty-five million years ago was too severe for natural selection to overcome. Mayr says, "The existing genetic organization of an animal or plant sets severe limits to its further evolution." Hence, the dinosaurs didn't evolve their way out of the dilemma, plausibly due to the limitation of their genotype.

The reptilian genotype began structuring 330 million years ago and was solidly established when a severe alteration of the environment initiated the dinosaurs decline. Climate change can be mild, moderate or severe and some animals can bear drastic changes better than others. Obviously, rapidly restructuring the dinosaurian genotype for a phenotype that would tolerate the catastrophe that wiped them out was not in the cards Nature held in her hands.

Concepts of rapid and gradual are applicable to evolutionary events whose pace is usually, but not always, guided by the rate of environmental change. Rapid human speciation from Pliocene apes was ten times faster than horses evolved from their ancestors. After the prompt emergence of hominids from the ancestral ape, the fossil record denotes the Australopiths and Homo hominids arose quickly at different times from their precursors. A famous descendant of the Pliocene apes, Homo erectus, was stationary in the first part of its existence.

The stasis ended and the pace stepped up when the Erectus brain started to expand. Judging by the inconsistent expansion denotes Erectus evolved in spurts, unless the spurts were gaps in the fossil record. The brain of Homo sapiens, the species described as wise, has remained static in the fossil record since it emerged. So did other hominids until they evolved into their descendants or faded out. A complete fossil record would illuminate the rate of hominid evolution, as a complete fossil record would do for other animals.

A fast evolutionary pace of an animal species is exemplified by a segment of its parent population budding off into a new adaptive zone like moving from a warm to a cool environment. The animals in budded segments change in degrees correlated with the difference in their new environment, and may become varieties or new species. They wouldn't change appreciably if they budded into a similar setting. On the other hand, animals moving into an extremely different environment, or the environment moving in on them, would have to hop to it and acclimate quickly by making changes needed to survive.

Provided the segment's parent population is static and not undergoing changes, it is in genetic equilibrium. This requires special conditions that are: the population must be large, all members breed randomly. There can be no natural selection, mutations or migration. When the conditions are upheld, the genes being exchanged, or gene frequencies will remain stable. The conditions are called the Hardy–Weinberg rule, which is an ideal model. In reality, the conditions don't hold true at the same time, or evolutionary events wouldn't occur. Even static species are undergoing some evolution, though it is negligible.

One violation of the rule occurs when some members of a small population don't have a chance to breed. When these members have genes that other members don't have, the genes may be lost from the population. This evolutionary mode is known as genetic drift and it operates in a small gene pool that has fewer genes to exchange. Genetic drift can initiate evolution without natural selection.

However, natural selection will eventually act on individuals and their new genetic composition. That is, those with the best traits for the environment will have a better chance to survive and reproduce. Genetic drift may lead to a mild or profound genetic restructuring that produces varieties or new species. A rapid and profound genetic restructuring of some apes through genetic drift, for instance, could have produced the first hominids with their profound restructured legs. Then again, a mutation would do the same if it caused the apes to move out of their lofty neighborhood.

Animals that violate the Hardy–Weinberg rule by migrating from the parent population to another locale are called founders. When the founders move they may form a bottleneck, which is called the founder effect. Founders usually wind up isolated in an outlying area that generally varies mildly from their previous location, but the variation can be severe. Whatever the variation, natural selection acts to accommodate them to their new home. Mayr says, a founder "population evolving in isolation is called an incipient (beginning) species."

Incipient species resemble hybrids like mules and ligers or beefalos, crosses between cows and buffalos. Incipient species can be called varieties, but they are progeny of one species and not two different species. Sometimes incipient species return to the parent population and interbreed, which causes them to lose their incipient status. The isolated incipient species may evolve into new or closely related species or become extinct. Meeting extinction is more likely when genes for a critical trait don't tag along in the bottleneck or the right mutations don't crop up. Evidently, longevity may be in inherited or mutated genes.

Founder populations unfold from geographic events like the flooding of the Bering Strait that separated Russia and Alaska at the end of the Pleistocene Epoch. A similar situation occurred in the same epoch when dry spells reduced a tropical rain forest to small areas. The forest's larger populations broke up into smaller populations that evolved into varieties and new species. Moreover, the founder effect may result from increases in population size causing splinter groups to disperse over large areas. Geographic speciation is the most common evolutionary mode, which Mayr says, "is the mode that has been thoroughly investigated."

Although incipient founders may resemble hybrids, the genetic structure of hybrids and founders is different. The chromosome number of two similar species such as horses and donkeys can be different. When they breed and conceive, their genetic material fuses and their chromosomes may misalign. Consequently, sterility results when their child, a hybrid mule manufactures sperms or eggs. Provided Homo sapiens and Neanderthals interbred and were near enough genetically, their offspring would have been fertile. A theory holds this happened and Homo sapiens pleasurably absorbed

Neanderthals. Thus, they didn't vanish, but were saved by sex.

ARTIFICIAL SELECTION

Father

Mother

Hybrid mule

CHAPTER 6

A human and chimpanzee hybrid called a humanzee is unknown unless the two species interbred behind closed laboratory doors. Artificially selected hybrids are mainly commercial products and some are used for scientific experiments. Genetic theory maintains a humanzee could be produced easily under controlled conditions. "This interbreeding may well be possible," Gould says, "so small are the genetic distances that separate us." Credibly, a humanzee would have more intelligence than a chimp and less than a person. The hybrid might tell what it's like to be a chimp, but its birth would certainly confound fundamentalist sects.

Commercially bred hybrids are preferred over their parents, a condition called hybrid vigor. Blight resistant vegetables, for example, are hardy hybrids produced by artificial selection. Darwin observed that pigeons with certain traits were bred for useful hybrids, but he had no knowledge of the genetics underlying the process. All dog breeds are the same species and products of artificial selection that reproduces fertile offspring. The effectiveness of artificial selection acting in a short time produces Saint Bernard and Greyhound dogs while Blue Whales and porpoises evolved over millions of years by the power of natural selection.

Whales and dogs are mammals, the animal class that includes cows, mice, goats and people. Their phenotypes, or bodies, are built from the DNA recipe of the mammalian genotype. Mammals have characteristics not found in other animals such as mammary glands that produce milk. The genes that code, or direct the construction of the glands are part of the mammalian genotype. Since reptiles and amphibians didn't select mammary glands is why snakes and frogs don't produce milk. Mammary genes and other genes specific to mammals were added to the reptilian genotype during the evolution of mammals from reptiles.

All genotypes were built on the genotype that was assembled for bacteria in the Precambrian seas. Later, the Cambrian Explosion produced the genotypes of thirty phyla of invertebrates and a subphylum of vertebrates wherein brewed the rudiments of the human genotype. Indeed, the explosion presented natural selection copious prospects to fabricate more kingdoms. "Most of the major phyla of invertebrates," Gould says, "made their appearance within a short span of a few million years."

The arthropod phylum is the largest, which includes lobsters, bumblebees and tarantulas. Lobsters and tarantulas are not social animals, but loners that seek nourishment on their own. Conversely, honeybees selected a cooperative social organization somewhere along their evolutionary path. Perhaps two transitional insects between bees and their precursor established a symbiotic relationship that evolved into a complex society. Social insects and mammals work together within their species for the common good. One works on instinct and the other is mindful. The social animals cooperate like the all for one and one for all *Three Musketeers*, by Alexander Dumas.

Characteristics of the social busy bees are exterior skeletons and joints. The exoskeleton of these arthropods limits their growth to no larger than a big crab. Some mollusks are soft-bodied animals such as octopuses and giant squid that can grow larger than fifty feet. Other mollusks such as snails and oysters have a hard outer shell. Some scientists think arthropods are dead ends, or at least limited in more evolution. Endoskeletons allow the growth of huge animals such as elephants and whales. Humans are

are smaller than the former, but larger than most animals.

Another major animal phylum that appeared in the Cambrian Period is classed as echinoderms. The phylum included starfish, sea cucumbers, sand dollars and an unidentified human grandparent. Echinoderms gave rise to a vertebrate subphylum represented by fish, amphibians, reptiles and mammals. Echinoderms and arthropods took separate paths by selecting different kinds of embryonic cell division.

Echinoderm and arthropod cells divided along different planes in the first stages of their embryo's growth. This was a random what if incident that was crucial to human evolution. All prospects for people would have stopped with echinoderms had they not selected their particular type of embryonic cell division. Echinoderms were essential for people who believe they are the end product of a planned progression. Echinoderms had to evolve or people might have taken the arthropod route and look like big bumblebees.

Two Starfish

Human Grandparents?

Mom and Pop

Nature worked six hundred million years on echinoderms and their descendants, selecting new genes while co-opting old genes to make people. After all, chimps weren't going to discover penicillin to save lives or invent atomic bombs to squander lives. Earlier, natural selection made bacteria from scratch followed by other human ancestors. To continue, Nature put chordates together from an unspecified progenitor. Chordates are animals identified by their notochord, a stiff rod lying beneath the nerve cord. The cord was modified in 100 million years into a backbone, one that tends to crumble in people, especially if the biped is overweight.

The acorn worm, lancelet and tunicate are chordates in one system of classification. They represent ancestral chordates of the Cambrian Period in transition between invertebrates and vertebrates. The three chordates are degenerates, or 'leftovers', inaccurate labels for animals like snakes and frogs that are reminders of the heyday of those cold-blooded creatures. Many creatures such as parasites that are accused of being degenerates or leftovers appeared recently on the geologic clock.

Frog

'Leftover' from the Age of Amphibians

Chordates arose and thrived as Nature added genes to the genotypes of one lineage of chordates and turned their notochords into backbones. Notochords and other parts modified into descendent parts with the same or different function are labeled preadaptions by disciples of intelligent design. The disciples believe the notochord was preplanned to make the human backbone. Still, had the troublesome backbone been a preadaption it wasn't quality controlled. What's more, intelligent designers might be loath to acknowledge the chimpanzee brain is a preadaption since the human brain is a lopsided version of the chimp brain.

Scientists who think evolution is random discount preadaptions that are predestined parts. They call parts like notochords exaptations saying they are made with available genes for the moment. Depending on viewpoints, lancelets are chordates like other potential grandparents have parts that are either preadaptations or exaptations. One part is a pharynx with a system of gill slits used to feed on small particles. The slits are assumed precursors of oxygen absorbing gills. The adult lancelet is the best example of a chordate with all the chordate characteristics, but that doesn't mean it was destined to become a new species.

Another chordate is the acorn worm that digs U shaped holes in the sea floor with an extendable muscular proboscis. The worm has a trace of a nerve cord and a cartilaginous structure analogous to the backbone of vertebrate animals. The similarity of the larval stages of acorn worms and echinoderms suggests the two animals share a common ancestor. This being so, denotes the early echinoderms and acorn worms were in a genealogical line of human grandparents. Still, many scientists say acorn worms were not human grandparents, but primeval uncles and aunts.

The likely human grandparents according to a popular theory were early tunicates. They are chordates called sea squirts because they squirt water when disturbed. The theory poses tunicates were forebears of a marine animal leading to fish, amphibians, reptiles and mammals, which includes people. Most sea squirts are vase-shaped sessile animals with highly developed swimming larvae that are similar to young fish. The larva resembles a tadpole and has a 'brain' of clustered nerve cells. Perhaps the nerve cluster is the prototype of the huge brain designed by an unknown entity or swollen by an uncommon occurrence.

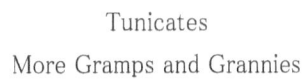

Tunicates
More Gramps and Grannies

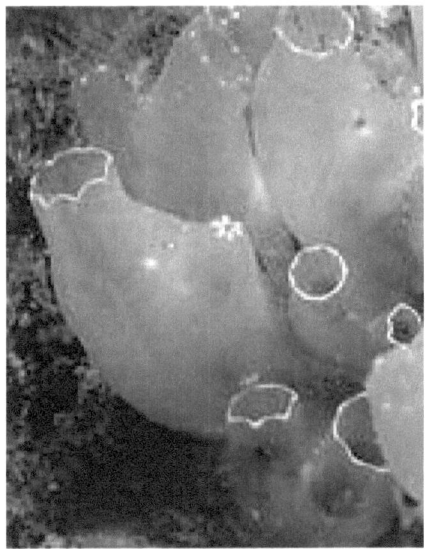

The tunicate larva breathes by gills and has a dorsal, or backside nerve cord enclosed in a notochord, precursor of the vertebrate spinal column. The larva and adult tunicate have virtually the same genotype, but like so many primitive aquatic animals, the larva and adult don't resemble each other. Two other animals with virtually the same genotype are chimpanzees and humans. They exhibit very similar behavioral patterns, but are dissimilar externally, analogous to the tunicate larva and adult.

Internally, the adult tunicate has a circulatory system with a heart and complete digestive system. Tunicates extract oxygen from water with a netlike sac, another possible preexisting part for gills and lungs. The tunicate is a good choice for the precursor of fish and humans concurring with a theory that the tunicate larva didn't acquire all the adult chordate traits. Instead, the larva became sexually mature and evolved into a fishlike vertebrate.

This being so, the ancestral tunicate of the Cambrian led to animals with expanding brains, one which shows remorse for decimating its fellow species. Sharing the blame for the decimator is the ostracoderm, a tunicate descendant and first fish to set fin on a lower rung of the evolutionary ladder. These ancestral grandparents were extinct jawless fish dating from the Ordovician Period. Ostracoderms were encased in scales and bony plates and it's thought the skeleton was cartilaginous. Some were two feet long, but most were smaller.

Ostracoderm

Old Granddad?

Ostracoderms were similar to the extant hagfish and lamprey, two primitive jawless fish with bodies like eels. Hagfish and lamprey skeletons are cartilaginous, and though the fish are placed higher on the animal tree, they have commonalities with the early chordates. Hagfish eggs hatch into small hagfish, whereas the lamprey spawns larvae that are unlike the adults. Most hagfish and lampreys are parasites with 'degenerate' bodies. Mayr poses parasites adapted their life style "because most evolutionary changes are dictated by the need to cope with current temporary changes of the physical and biotic environment."

Parasites such as tapeworms are 'lower' organisms that are well represented in the biological kingdoms. Tapeworms hook their mouth to the intestines of their host and suck out nutrients. Repulsive to people, the worm's means of subsistence discredits the notion of evolutionary progress. Parasites dumped more parts than they added, which gives the false impression they reversed evolution by climbing down the Tree of Life. Progress and reversal simply means adding or subtracting parts or intensifying or reducing their function. Parasites show evolution is as nonprogressive as extinction and their 'degenerate' phenotype is another indicator that random chance drives evolution.

Tapeworm

Disgusting Cousin

Needs no straw.

Degenerate jawless fish have a linkage outside the fossil record on the molecular level that connects them to their human and chimpanzee descendants. Comparing the chemistries of the hagfish and lamprey with chimpanzee and human chemistries reveals their kinship through the antibody immunoglobulin (Ig), a protein. Chiarelli says, "The most recent species in an evolutionary sense in which the Ig (immunoglobulin) molecules have been found are the hagfish and lamprey.

"In some non-human primates the serum contains the same kind of molecules as in Man. For example, in chimpanzees certain chains of the polypeptide are very similar to the corresponding human IG molecules." Though people have come a long way from the Ordovician fish without a jaw in its head, the polypeptide apple still falls from the same molecular tree. The odds get better and better that people had some grandparents that existed by parasitism.

The parasitic hagfish and lampreys are models for the ancestral jawless fish that diverged into numerous intermediate animals that gave rise to new species and more intermediates in the Silurian Period. Somewhere in that mass was a human predecessor that would grow little by little on branches of the animal bush until it became men, women and children. The figurative grandparent would evolve through amphibians, reptiles and mammals while acquiring major adaptations so people could breathe air, circulate warm blood, walk erect and grow bigger brains.

Diverging from the jawless ostracoderms in the Devonian Period was a bony fish and next human ancestor. Ahead, there was ample time and shifting environments to fashion the genotypes of frogs, snakes, kangaroos, gorillas and people. The Devonian Period is known as the Age of Fish when the emergence and proliferation of new species of fishes burgeoned. Nature worked 24/7 her wonders to perform and constructed multitudinous animals that couldn't appreciate the difference between a smelly swamp and the beauty of a rainbow--or the suffering of others.

The Devonian produced three major fish lineages represented today by the ray-finned fish and the primitive leftovers, lungfish and coelacanths. Ray-finned fishes comprise the familiar species from cartilaginous sharks to bony salmon. Lungfish are eel shaped and live in poorly oxygenated waters, breathing with a specialized lung that is its main respiratory organ. Coelacanths are 'dead-ends' that live in deep waters and have lobes, or rounded protrusions on the end of their fins. The only remaining coelacanth species is called a living fossil because it has not changed in millions of years.

The theoretical ancestor of amphibians was a primitive Devonian fish with paired air sacs that inflated or deflated to raise or lower it in the water. Two of its divergent lineages gave rise to existing bony fish and lungfish. As the intermediate fishes that would lead to the lungfish and bony fish radiated into different niches, their air sacs slowly altered to adjust to their surroundings. The air sacs continued to modify until they became swim bladders in the bony fish and lungs in the lungfish. The microevolution of swim bladders into lungs paved the way for people.

A supposition for the evolution of lungfish is a transitional creature arose from fishes that lived in an inland body of water whose oxygen content was decreasing. As evolution intensified in the isolated population, some became intermediates, varieties and new species. Natural selection built on their preexisting parts of one or more emerging lungfish that folded the air sacs inward. The enfolding increased the surface area of the waxing lungs and facilitated the absorption of oxygen. Later, lots of folks would give their waxed lungs a hard time by mixing the oxygen with lots of smoke.

Lungfish are progeny of those Devonian fish that survived bad times by breathing air while remaining in the water or mud. However, the fish that was to turn into amphibians in two hundred million years and into humans four hundred million years after that needed more than lungs. The common ancestor of frogs and people needed lobes to help it evolve into a four-legged animal, or tetrapod. Obviously, it took lots of exaptations or preadaptations along the way to make Homo sapiens.

Since the fossil record is sketchy, the traits of the fish to be amphibians are inferred from available fossils and living coelacanths and lungfish. There are several speculative lines of divergence for the precursor fish. One line is, the common ancestor of coelacanths and lungfish evolved from some primitive fishes that were marooned in inland waters that were receding or stagnant or both.

The common ancestor diverged in two lines of transitional fish and one line give rise to coelacanths and the other to lungfish. Another opinion is, some precursor lungfish lived in slowly drying environs and over uncountable years adapted lobes to trek to lakes or rivers. Perhaps the lobes were already present in the precursor fish, having been modified earlier from fins to slink on or for another reason. The lobes were either preadaptions for a decreed task or exaptations for incidental events.

Comparing mitochondrial DNA of existing amphibians and lungfish implies that lungfish are the closest relatives of amphibians and the early lungfish was a human ancestor. This being so, it followers that the nimble fingers of a virtuoso evoking rhapsodies from a violin and the lithe limbs of a graceful ballerina gliding across the stage were coded by variations or the same genes that once coded for the lobes of a lungfish groveling out of the muck.

Lungfish Circa 400,000,000 BC

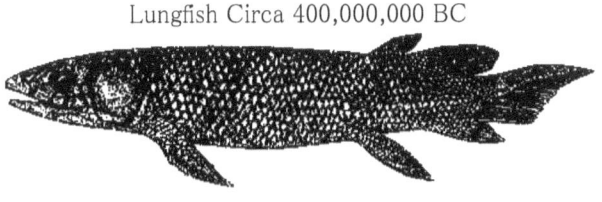

Human grandparent

Scientists don't identify the common ancestor of lungfish and coelacanths or specify what mode initiated their evolution. The common mode is natural selection, and the two fish evolved over millions of years. A ten million–year evolution of a species represents 1/450 of earth's history, and a half million years is a geologic moment. The five million years of human evolution is 1/900 of earth's history that rapidly produced a puzzling primate with an almost bald body. Further, the substandard body parts that plague the puzzling primate suggests natural selection didn't initiate the first, or some subsequent events of human evolution.

Deficient traits that perform poorly, out of numerous traits offered by genetic variation emphasize that chance determines those traits. Since natural selection favors individuals in a population, deficient traits stresses the best of the worst phenotypes are selected, if the best aren't offered. Animals like the duckbill platypus indicate Nature is limited and can choose only available phenotypes with their tagalong genes, which doesn't support the beliefs of intelligent designers. The phenotypes of platypuses and people question if natural selection was involved when those two oddities or one of their precursors began to evolve.

Duckbill Platypus

Origin as quizzical as human origin

An animal deviating from the norm followed the proliferation of fishes in the Devonian seas and inland waters. A theory holds that the oceans were overpopulated and natural selection acted to modify a fish to emigrate before the congestion became critical. Solving the problem of overcrowding with fleeing fishes is a notion attuned to intelligent design. Surely, to continue the process of making humans, the big plan contained instructions how to get a people pioneer out of the water and onto land.

Evolutionists wouldn't acknowledge a big plan because the amphibians that followed the fish deviated by not adapting completely to water or land. This would slow down making people unless something thought half the fun was getting there. Whatever happened, the awkward amphibians flourished for nearly one hundred million years and today exist as degenerates that are struggling in many polluted habitats. Amphibians haven't adapted to salt water, which could be their last refuge as pollution continues. Adapting to salt water seems as doable as a lineage of reptiles learning to fly, but the amphibian genotype might not be sufficiently flexible.

The genotypes of ancestral fish, amphibians, reptiles and mammals were built upon the genotypes of their predecessors starting with Precambrian bacteria. The genotypes of animals dwelling in the early waters coded for phenotypes fitted with swimming structures such as flagella or fins. The animals breathed by diffusion or gills and eventually some of their descendants evolved into new species by learning to creep and gulp air. Later, a few learned to walk and breathe by 'inventing' limbs and lungs. The physical, chemical and behavioral instincts incorporated in their genotype became fundamental to the chordate genotype.

Genes coding for the characteristics of chordates were handed down to the evolving vertebrate phylum. The chordate characters appear in vertebrate embryos and develop in some, but not all adults. Gill slits occur in all vertebrate embryos including humans, but only become gills in fish and amphibian tadpoles. Mayr says an early human embryo is "very similar not only to embryos of other mammals (dog, cow, mouse), but in its early stages even to those of reptiles, amphibians, and fishes." The appearance and loss of primitive characters in a mammal, bird or reptile embryo is called recapitulation.

An old theory proposes embryos repeat, or recapitulate, the embryonic stages of their progenitors. Reptiles, birds and mammals grow from embryos resembling fish embryos with gill slits. However, they don't use gills, which questions why natural selection didn't eliminate unneeded characters. Mayr says the characters serve "as embryonic 'organizers' in the ensuing steps of development." Conversely, other scientists say recapitulation isn't understood. The theory is explored in a sci-fi movie about an English nobleman whose embryo stopped developing at the stage of an amphibian. The sad fellow was born with a frog's body and lived in shame.

Gill slits that become gills in amphibian tadpoles and fish have been modified through the ages into other structures in the necks of reptiles, birds and mammals. Some reptilian parts that were modified in mammals can be observed in an accurate section of the fossil record of the reptilian lineage leading to mammals. Reptilian embryos, for example, produce bones that natural selection altered for another purpose in mammals. The bones in the intermediate animals were functional and therefore did not remain in useless limbo while being adjusted for a new role.

"Only one bone, called the dentary, builds the mammalian jaw," Gould says, "while reptiles retain several small bones in the rear portion of the jaw. We can trace through a lovely sequence of intermediates, the reduction of these small reptilian bones, and their eventual disappearance of exclusion from the jaw, including the remarkable passage of the reptilian articulation bones into the mammalian middle ear (where they became our malleus and incus, or hammer and anvil). The transitional species maintains a double jaw joint, with both the old articulation of reptiles and the new connection of mammals already in place!

Thus, one joint could be lost, with passage of its bones into the ear, while the other articulation continued to guarantee a properly hinged jaw." This being so, a snake's jawbones that expand to swallow a rat are modifications in human and chimpanzee auditory structures. Hence, mammalian jawbones and ears started with the 'preadaptive' or 'exaptative' gill slits of a fish. Meanwhile, animals that evolved from the fishy human ancestor became bushes of intermediates, new species, offshoots, dead-ends or they met extinction. One stem on a bush grew into amphibians.

A sprig on the bush was a fishlike amphibian that inhabited fresh water. Its classification was inferred from a 365 million year-old fossil foreleg bone dated in the late Devonian. The bone belonged to a two and a half foot amphibious-like transitional animal called a robust salamander. Its powerful foreleg connected to its shoulder by a hinge instead of the ball and socket joint of mammals. The limb was fixed in a sprawling stance like an alligator's foreleg that stabilizes its body in a current. The bone's configuration indicates the animal lived in streams where it broached fast currents.

The amphibious-like specimen suggests that many features selected for land evolved while fish lived in water. Fish tailoring their bodies for land was a sound strategy since plants had moved there in the Silurian and were ready for the taking. More good news was the fine dining on nutritious marine arthropods that were previously modified to radiate and exploit terrestrial niches. The arthropods that migrated onto land such as insects soon became prey in the food chain. The terrestrial descendants of the robust salamander feasted on a banquet of fresh plants and animals.

The robust salamander was in theory a human grandparent in the series with lungfish. To preserve the series and get a human pioneer on land, intelligent design needed to move some vertebrates on land and turn one lineage into amphibians. A quicker and easier route would have been to put a few echinoderms on the ground earlier so they could give rise to vertebrates. That route wasn't taken, which was either poor planning or natural selection was determined to take any route available. Nature stuck with the robust salamander, and through transitional species it gave rise to amphibians.

CHAPTER 7

While the intermediate animals converted into amphibians, they became semi-adapted to land. They munched plants and zapped bugs with their long sticky tongues. Natural selection favored body parts that let them walk and breathe with lungs and through their skin where oxygen and carbon dioxide diffused. Skin respiration allowed them to remain longer under water and hibernate or aestivate during cold or hot weather. Their backbones altered to shift weight onto their wide-straddled limbs that were positioned at right angles to the ground to reduce stress on the spinal cord and back. No such luck for their feeble bipedal descendants.

Adult amphibians swapped their tadpole gills for lungs and added a third auricle to their heart. Biochemical, physiological, physical and innate behavioral characteristics altered to adjust them to the new environs. "The mixture of fishlike characteristics with those of the tetrapods," Beerbower says, "demonstrates that these amphibians were near the origin of the tetrapods." They were human bound, but they had more projects to complete, which they did over millions of years.

Amphibians retained the reproductive ways of fish and spawned in water where the male discharged sperm over the gelatinous eggs the female released. Their 'baby' tadpoles hatched in water and breathed with gills like tiny fish. Soon, they metamorphosed or changed into aquatic or terrestrial adults. Terrestrial amphibians were limited in the distance they ranged from their birthing waters since they had to return there for sex. Many amphibian "lineages that returned entirely to aquatic habitats demonstrates the uncertainty of their terrestrial adaptations," Beerbower says. Nature fitted them for land, but kept then watered down in their conjugal rendezvous.

Once amphibians were established, they thrived and reached their zenith during the Age of Amphibians in the Carboniferous Period. Primeval amphibians lived in terrestrial and aquatic environs while some resided in both places. Their miscellaneous bodies were adapted for the diverse habitats they exploited. Amphibians varied from tiny frog–like creatures to brutes with big teeth and armor–covered bodies as huge as grizzly bears. Amphibians had their heyday, but their reproductive constraints gave them the role of big frogs in small ponds. The role started changing in the late Carboniferous when a minor linage began diverging in a reptilian direction

Artistic imagery of brutish amphibians in the Carboniferous Period showing an unlucky time traveler.

One hypothesis for the amphibians diverging poses their spawning waters were crowded; putting them in the same overloaded boat abandoned by their fishy precursors. The congestion prompted an exodus of the minor lineage to a promising land. The adaptations selected during the journey produced amphibious–reptilian intermediates. Their instinct urged them to exploit new environs and radiate into various locales. Perhaps a population segment moved through a bottleneck into an isolated area where evolution is intense and they were turned into reptiles. Whatever transpired, consensus is natural selection was the agent that transformed amphibians into reptiles and germinated the reptilian bush.

The reptilian bush flourished in the Jurassic Period and Nature was near completing the people project, or human evolution. Intelligent designers believe the project began when the earth formed, but differ with evolutionists on the timeline. Regardless, mammal precursors of people were just around the geologic corner. Maintaining the pace, reptiles upgraded the three-chambered amphibian heart with a partition in the main pumping chamber that prevented poorer oxygenated blood from mixing with richer blood. Reptiles repositioned their legs closer under their bodies and could move faster and farther away from the ponds.

Their lungs were larger than amphibian lungs and their skin wasn't used for diffusion or breathing. Tough scales protected them from predators and drying out in hot environments such as deserts. Reptilian eggs were drought resistant and could be hatched to populate barren areas where no vertebrate had gone before. At last they didn't 'worry' about getting back to a pond for pleasure. Many other physical and biochemical adaptations enhanced their progress toward fully terrestrial animals. Clearly, their progress seems goal oriented, but evolutionists maintain a progression from simple to complex animals is only a series of random events.

The rise of reptiles from amphibians was sustained by multitudes of genes resulting from chance events. The amphibious-reptilian intermediates forsook their tadpoles and drastically overhauled their gelatinous eggs. Sticking with gooey eggs would forestall reptilian evolution and keep human ancestors lounging around the water, hopping in solely at the innate urge to copulate. Acting over millions of years, the diverging amphibians remodeled their waterlogged egg into the dryer amniotic egg, exemplified by a turtle and chicken egg. The reptile eggshell has a leathery texture whereas their descendent offshoots, the birds, have hard shells made of calcium compounds.

During amphibians transition into reptiles the fertilized egg became enclosed in a leathery shell. Internal conception occurred by the male passing sperm through his cloaca into the female cloaca. However, reptilian awareness was insufficient for the copulating couple to appreciate the act. Later, a huge brain would make more ado about sex than instinctive reproduction. Cloacas are present in amphibians, reptiles, birds, most fishes and some invertebrates. The genital, urinary and intestinal tracts open into the cloaca that gave rise to the erectile penis of mammals. The less dutiful penis has only two chores: expel urine and deliver sperm.

Cloaca coding genes, along with other amphibian genes were added to the reptilian genotype. Some amphibian characteristics aren't present in reptiles and those unused genes might be stored with other genes with no apparent purpose. Reptilian systems were changing to meet the challenges of new adaptive zones. Their reproductive strategy is more 'advanced' than amphibian strategy, but less than in their mammalian descendants. Reptilian eggs are essentially portable wombs that the mother lays on land unattended. The hatchlings are replicas of adults that must face the vicissitudes of the cruel world to which many succumb.

What's more, the mother or unknown father of some reptiles might happen upon their growing children and cannibalize the little tots. Unlike people, those reptiles could "chews" their family as well as their friends. Further, and similar to parental care of mammals, it's hypothesized a version of motherly love existed in a primeval Jurassic park.

The hypothesis poses a dinosaurian social structure existed wherein female parents watched over their eggs and attended to their young. Certainly, parental care in some dinosaur herds would have helped the children through their awkward age so they weren't eaten alive.

Conceivably, primitive reptilian social groups seeded the emotional properties that sustain egalitarian societies of mammals such as those of social chimpanzees and elephants. Anthropologists define the emotional properties broadly as love, which encompasses the altruistic behaviors that support the human egalitarian social organization. Perhaps human love evolved from several behaviors of reptilian precursors whose smaller brains could retain a kernel of learned information. Concurring with the theoretical Baldwin effect, a learned behavior may be inherited by selecting other traits that reinforce the behavior. Dinosaurs extending paternal care could be a survival mechanism of cumulative learning Baldwinized into the genotype.

The Baldwin effect does not support Lamarckism, a theory purporting that acquired traits are heritable. According to Lamarckism, the necks of giraffes lengthened over time by giraffes stretching their heads into treetops to eat leaves. "The pathway from nucleic acids to proteins is a one-way street." Mayr says. "Proteins and information contained in them cannot be translated back into nucleic acids." That is, DNA builds proteins, but proteins can't build DNA, so the inheritance of acquired characters is impossible. The enlarged muscles of a weightlifter can't be translated into the weightlifter's genotype and be passed on to the children.

Heritable traits such as eye color and their coding genes follow the rules of Mendelian genetics. Genes are dominant, co-dominant and recessive, which, with other factors build the characters of an individual. A new generation of sexual organisms begins in the reproductive organs with the intricate processes of chromosomal rearrangement, meiosis, crossing-over and mutations. Each animal is genetically unique, and those with the best traits have the best chance to survive and reproduce. The fastest gazelle, for example, that outruns a predator is chosen with its genes by natural selection over a slow one that becomes a meal.

Individual differences among sexually reproducing animals like people are prepared for natural selection's scrutiny when a sperm and ovum unite. The number of gene combinations available at conception is astronomical, which is the reason each person is genetically distinct. The traits are random results and may be good, fair, poor or rock bottom. Natural selection can't predetermine the quality of traits, but can discriminate and pick the most useful ones. People, and supposedly other Homo hominids, survived with some poor and rock bottom traits. Expanding brains compensated for the deficient traits, else the Homo lineage might have become extinct.

A good trait is hypothetically illustrated in a fish population that adapted a strategy to spring from a hole and catch passing prey fish. Poorly adapted fish had shorter lives and less offspring than well-adapted fish. Examples of good traits in a different animal are blue eyes and light skin that replaced the brown eyes and dark skin of early humans leaving Africa. Apparently, lighter colored eyes and skin were adapted in response to variations in sunlight. Supposition holds the lighter colors were coded by mutations favored by natural selection.

Efficiently modifying animals seems designed, considering "proper adaptions are illustrated by the way a lioness hunts, a horse runs, and a hippo wallows," Gould says. Their adaptions "intimate natural selection cast them precisely for their appointed roles." Refuting any casting or precise design are odd animals like the duckbill platypus. "The platypus, has hair and suckles its young with milk and has other characteristics of primitive mammals," Mayr says, "but lays eggs, like reptiles, and has some "dead-end" specializations, like a poison spur and a duckbill." Clearly, platypuses and other oddities emphasize that natural selection doesn't design organisms.

More in tune with design is mosaic evolution, a common mode that modifies only the parts needed for changing conditions. "An organism is a carefully balanced, harmonious system, no part which can change without having an effect on other parts," Mayr says. "Let us consider the increase in the size of teeth in horses. This change requires a larger jaw, and in turn a larger skull." Over thousands of generations, Nature ground out bigger choppers to munch new grasses that arose during altering environments. Meanwhile, she enlarged the skulls and overhauled the bodies of the evolving horses.

Other choppers the mosaic mode supposedly enlarged were the teeth of Saber Tooth tigers. The tigers were in an 'arms race' with large prey animals whose hides were being thickened by mosaic evolution. Perhaps intelligent design made a smart move and enlarged the tiger's teeth. Yet, it's been proposed that the tiger became extinct because the teeth grew too big to be effective and the tiger adapted itself into extinction. Continually adapting a part until it gets too big for its britches seems illogical. Credibly, the conviction of many scientists is correct, that Nature doesn't adapt organisms into extinction.

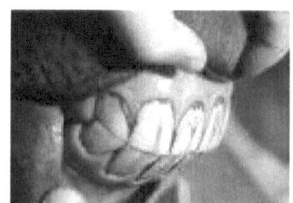

Horse teeth

Mosaic evolution of teeth

What big teeth you have grandma

The tigers might have vanished because their prey became extinct and took the big cats with them. Also possible is a rapid environmental change didn't allow Nature to adjust the tiger's teeth. A fast environmental change is often proposed for mammoths vanishing and not because their tusks got so big they drove the big fellows to extinction. A good bet is a predator wielding spears did them in, a common practice of the two-legged species. Also, it's possible the death rate of the tigers and mammoths exceeded the birth rate for an unspecified reason and they died out.

Declining birth rates initiated by competition with other animals are frequently proposed for the source of extinction. Reptiles, for instance, didn't eliminate their amphibian 'parents', but demoted them to the ill-defined category of degenerates. Reptiles evolved from the minor lineage of amphibians in eighty million years, and once the transitional animals lost their amphibious character, they emerged as full-fledged reptiles. The reptilian world was fruited with great expanses of forested trees and plains of emerald grasses. The bountiful herbage promoted their growth and they became the largest of all land animals.

The burly dinosaurs dominated the earth during the Age of Reptiles in the Mesozoic Era. Tyrannosaurus Rex was a large predacious reptile forty feet long and weighed eleven tons.

That past jumbo is a sci-fi favorite that gobbles up Neanderthal bumpkins who didn't exist then and weren't bumpkins in the first place. Brontosaurus was a whopping eighty-ton plant eater four stories tall and bigger than ten elephants. The cold-blooded rulers abdicated their throne at the close of the Cretaceous Period and were inducted into the imaginary Hall of Extinction.

The dinosaurs decline is paraded in layers of earth containing their fossils that diminish and vanish in succeeding layers. The prevailing theory of their extinction proposes a massive meteorite struck the earth and disrupted the environment. The harmful conditions killed vast grasslands causing numerous herbivores to die off followed by their dependent predators. Another theory holds the brainier mammals ate their eggs and the next generation was digested before it hatched. Some scientists say the fossil layers indicate the dinosaurs were following diminishing vegetation that contained roughage and purgative oils.

Consequently, the lack of bulk and emetic oils affected the reproductive output of the herbivorous reptiles. Their poor diet is analogous to the health evaluation 'you are what you eat', or in their case, it's what they didn't eat. Other scientists believe pathogenic microbes caused an annihilating Black Death that drove their birth rate below their death rate. Natural selection is determined to salvage animals like the doomed dinosaurs from termination, but it has limitations. Mayr says nothing confirms its limitations "more convincingly than the fact that 99.99 or more percent of all evolutionary lines have become extinct."

Tyrannosaurus Rex

The bigger they are, the harder they fall

The Age of Reptiles began failing sixty-five million years ago at the start of the Tertiary Period. During the decline, the smaller mammals that were scurrying in the shadows increased in number and size. Mammals are characterized by a complex neural system and greater intelligence. Their bodies are covered with hair or fur and they are the only endothermic or warm-blooded animals, except for birds. Mammals have sweat glands and mammary glands with nipples that secrete milk, a mixture of water and nutrients. The duckbill platypus and spiny anteaters are a subclass of mammals that deviate from the norm.

The mammalian embryo develops from a fertilized egg in the uterus or womb, a self-contained incubator. Babies are born in degrees from near helplessness to virtual self-reliance. Mothers nurture and protect their youngsters from short to long periods, and in certain species, the fathers assist them. Some mother monkeys wean their children in a few weeks whereas an ape's childhood lasts several years and humans much longer. Human and ape mother's love is lasting and many human children live with their parents until somebody dies. An anthropological theory poses that extended adolescence is one result of a nonadaptive primate society.

Primate genotypes were built from the succeeding genotypes of fish, amphibians, reptiles and mammal classes. The succeeding animal classes 'advanced' by co-opting old and adding new genes to their genotypes. The human genotype was built at the end of the 'advancement' and is claimed by some intelligent designers to be an orderly work of sound planning. However, paramammals and marsupials that evolved before humans are two animals that question sound planning. The fossil record indicates marsupials arose in North America from paramammals in the late Cretaceous Period.

Paramammals appeared about the time the first reptiles evolved, 100 million years before the dinosaurs. They were possibly reptilian offshoots since they had mixed traits of reptiles and mammals. Paramammals aren't labeled one hundred percent this or that and might have been transitional animals. Sometimes paramammals with more reptile than mammal traits are called parareptiles. Another mishmash of traits is the often-cited duckbill platypus that has more mammalian than reptilian characteristics. The primitive animal is a zoological oddity that could pass for a mammal-like paramammal.

Paramammal
Circa 300,000,000 years ago

Paramammals existed until the end of the Permian Period when the largest of five major extinctions eliminated more than 50% of all species. The Permian Extinction is tied to the theory of plate tectonics that proposes the earth's surface is composed of rigid plates that creep around. The plates can travel five hundred to a thousand miles in ten million years, an eternity compared to a human lifetime. During the fifty million-year Permian Period the rambling plates could move thousands of miles, causing continental drift that breaks up landmasses situated on the plates.

The smaller land segments meander around the globe and bump into other detached sections of earth. The pileups produce geologic events such as the formation of oceans and continents, mountain building, earthquakes and volcanic eruption. Animals that can't adapt to new conditions generated by tectonic jumbling suffer and decline or vanish. Try as she may, Nature can do just so much to adapt animals to changing circumstances. As Mayr says, "There are definite limits to the effectiveness of selection." Clearly, bad news for all the animals people are running off the earth in the biggest extinction of all.

Other animals natural selection couldn't rescue in the Permian Extinction left empty habitats that reptiles and other radiating animals filled in the Triassic Period. The fossil record is unclear if paramammals disappeared or just declined. Provided they vanished signifies another animal gave rise to mammals, but consensus holds this is not the case. A popular theory poses the paramammals were intermediates and gave rise to mammals when reptiles dwindled. Chronologically, paramammals evolved through reptile-like to mammal-like intermediates and became complete mammals, the animal class graced with the cerebral primate.

The cerebral primate was fabricated in four billion years and it's puzzling why Nature wound up with such a frail body. Provided intelligent design built the nearly hairless biped, it encompassed four-tenths of the earth's predicted existence. When the earth's time is up on its ten billionth birthday, the sun will burn out and expand to the orbit of Mars. The world will fry to a cinder and the bloated sun will collapse into a black dwarf. Six billion years remain for people to do lots of mischief with lots of regret doing it unless they vanish in the meantime.

Humans were constructed on the genotypes of their precursors through reproduction that operates by cell division in bacteria to 'conventional' sex in mammals. Conversely, 'unconventional' hermaphrodites connect their male and female sexual structures to reproduce, a primal kind of bisexuality. Atypical sexual practices such as bisexuality and homosexuality agitate homophobes whose fear is induced by a response called pseudospeciation. The term means false species that allows people to see other people as a different species. Evidently, the apple falls near the tree because pseudospeciation was observed in disturbed chimpanzees and has been one curse of human society for 10,000 years.

People may mildly psuedospeciate curious animals that don't conform to the perception of normal. Curiosities are the most impressive and imply the only purpose of selection is to equip organisms to survive. "The best illustrations of adaptation by evolution are those that strike our intuition as peculiar or bizarre," Gould says. "Darwin understood this, and he focused on features that would be out of place in a world constructed by perfect wisdom." Homo sapiens and duckbill platypuses question perfect wisdom, as do atypical sexual practices, but in Nature nothing is wrong or right since she doesn't consider moral concepts.

Questioning perfect wisdom is the mammal subclass marsupials, whose female has two vaginas, each leading to a chamber of a two-chambered womb. The male sports a two-pronged penis he inserts into the twin vaginas like plugging an electric cord into a wall socket. One or two-prongs are reproductive selections, but intuitively the two-prong one seems peculiar. Yet, if men had a two-pronger, their reproductive partners would see it, or them, as standard equipment. The sexual organs of kangaroos and humans appear as coincidental adaptations for the same purpose.

Two-prong kangaroo penis

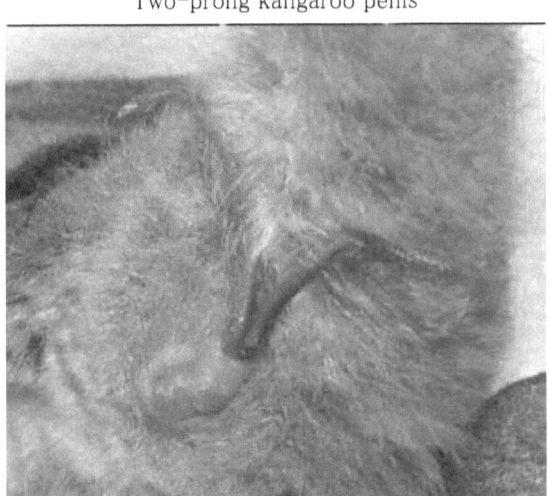

Adapted for twin vaginas

The four-prong penis of the male echidna or spiny anteater is another coincidental selection. Consensus is the anteater and its platypus cousin share a common ancestor with marsupials. This denotes as anteaters and marsupials diverged from the ancestor the two and four-prong penises were constructed from similar genetic variation. Deductively, Nature would have constructed a penis with the same number of prongs for both animals, except the genes available for the penises varied. Natural selection is determined to build an animal to survive and reproduce, but it cannot choose particular genes. Come what may is the dictate, and nothing more.

Spiny anteater. Kangaroo relative
and Platypus cousin

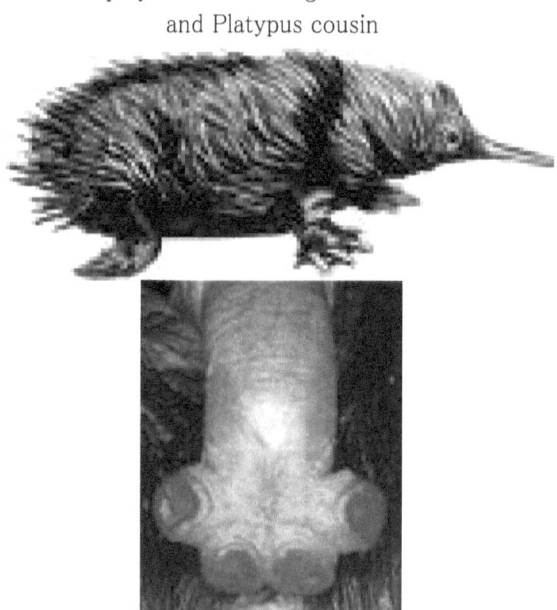

Anteater's four prong penis

   Coincidental penile selections and adapting different animals for similar habitats are distinct processes. Two more coincidental adaptations are one or two-hump camels and one or two-horn rhinoceroses. Conversely, marsupials evolved by convergent evolution, which selects similar traits for different animals evolving in similar environments. Tasmanian wolves and wombats are marsupials that resemble mammalian dogs and ground hogs. Bats are mammals that suckle their young while birds lay eggs and belong to the class Aves. Birds evolved from bipedal dinosaurs before bats. Birds and bats exemplify convergent evolution adapting animals in different classes to the air millions of years apart.

Two hump camel　　　　　　One hump camel

Coincidental adaptations that serve the same purpose

Two horn Rhinoceros　　　　　One horn Rhinoceros

CHAPTER 8

"Convergent evolution," Mayr says, "is a phenomenon that convincingly illustrates the power of natural selection. Convergence illustrates beautifully how selection is able to make use of the intrinsic variability of organisms to engineer adapted types for almost any kind of environmental niche." Convergent evolution may occur when efficiently adapted animals become extinct and are replaced by animals with similar physiques. The extinct Ichthyosaur, for instance, was an ocean reptile that resembled porpoises and dolphins, two marine animals. When the ocean reptiles died out, marine mammals succeeded them.

Convergent traits that different mammals selected are prolonged embryonic development and the nurturing placenta that lines the uterus. Instead of a placenta, a convergent trait of an expectant kangaroo is a yoke sac. The yoke's nutrients are absorbed by the embryo that later emerges in a rudimentary state. The minute marsupial crawls up its mother's body and into the abdominal pouch where it attaches on a nipple to suckle milk. The youngster is called a joey, and when it matures it moves to ground and may remain with the mother for a year.

Kangaroos adapted diapause, a reproductive method that postpones embryonic development. Diapause stops or reduces an activity and is a convergent adaption of other animals such as mammalian rodents. Kangaroo mothers utilize diapause and can mate when their immature newborn leaves the womb for the pouch. When the womb is vacated and an egg fertilized, the egg starts dividing to a point when diapause intervenes and stops cell division. When dividing ceases, the zygote remains arrested while the offspring grows in the pouch.

When joey leaves the pouch, or occasionally dies, the zygote resumes dividing and develops into a fetus. The fetus grows into the bantam creature that crawls into the pouch. Embryonic diapause enables three formative stages of kangaroos to exist. One is a zygote in suspended animation, another is a suckling in the pouch and a third is a joey under maternal care. Diapause and one or two prong penises were built by directions of their coding genes taken from the helter-skelter offering of genetic variation. Adaptions that intuitively seem odd emphasize the randomness of natural selection.

Mother Kangaroo with little joey

A curiosity to human perception

The peculiar kangaroo is nothing next to the tiny fly that lives and feeds on fungi, mainly mushrooms. The flies reproduce sexually until the offspring start feeding on the mushrooms where they switch-hit and reproduce asexually. Since the little weirdos lack a nourishing egg yoke, they selected to develop within their mother's tissues. To grow, they permeate her body and devour their mother from inside out. A few days later, the cannibals emerge from her lifeless body they have reduced to a hollow shell. Soon, the mother-munching matricides will have children of their own and in turn be gobbled up.

Another repulsive story is the mother beetle that gives birth to a single male larva. Shortly after the newborn emerges, the creep sticks his head into her private parts and devours his mom. Obviously, the ungrateful juvenile didn't have a father as a role model. Cannibalistic insects, of course, don't have the cerebral equipment to feel maternal or any kind of love. Cannibalism is part of that beetle's strategy to promote survival. Perhaps Nature was limited in available genes to adapt a 'conventional' mother and father reproductive process. There are numerous survival strategies in Nature that might offend human sensitivity.

Curious for sure, but entertaining instead of repulsive, are two female clams of closely related species. They live partly buried in bottom sediments with their rear ends sticking out. Gould says their larvae "can't develop without a free ride upon fishes during their early growth." The two closely related females devised gimmicks to attract fish that give the freeloading larvae a gratuitous journey. One female clam lures fish with a pigmented membrane she uses to "produce an unusual, rhythmic motion."

The other female cleverly builds a decoy that mimics a fish complete with an 'eye' and swimming motion. Deductively, natural selection used some preexisting parts of the clam's progenitor to counterfeit a fish to 'fool' real fishes. Devising gimmicks implies necessity served by chance will commandeer anything that is available. Intelligent designers believe adaptations such as fish faking are planned, even though these adaptations appear complicated and take indirect routes. Natural selection is not complicated and uses what is at hand, but completing time consuming adaptations seems inefficient

Natural selection fulfills its role by favoring the fittest individuals with their tagalong genes. Individual differences among animals in a population give Nature choices. The fastest running zebra, for instance, has the best chance to outrun a hungry lion and live longer to pass on its genes. Outrunning its fellow zebras in a race with a lion is an added plus for the fastest zebra. However, if the swift zebra trips and falls, its superior genes will be devoured in the unlucky accident. Zebra populations depend on natural selection to favor those zebras with 'fast' genes for the next generation.

Zebras and other animals are guarded by natural selection that Darwin says "is daily and hourly scrutinizing, throughout the world, every variation even the slightest." While scrutinizing, natural selection finds easy, difficult and impossible tasks. Picking speedy zebras is easy, whereas natural selection couldn't pressure and force the salvation of the diminishing dinosaurs. However, pressure or force "are strictly metaphorical, and that there is no such force or pressure connected with selection, as there is in discussions in the physical sciences," Mayr says. Natural selection promotes survival of the fittest by pressuring and forcing the weakest organisms out of existence.

Nature deals with small and large numbers of organisms to oversee the modes of evolution. Small events occur during microevolution at or below the species level. Through a small or micro event some zebras 'lost their stripes' and gave rise to quaggas, a subspecies of zebras without stripes. Quaggas evolved rapidly, conceivably due to disrupted gene flow in an isolated zebra population. However, they met extinction even quicker as victims of human hunters. Two other small events

were apes giving rise to hominids and chimpanzees to
bonobos, a chimp subspecies.

Last living Quagga

A typical result of human speciation

Circa 1883

Big events occur above the species level during
macroevolution such as amphibians giving rise to reptiles and
mammals evolving from reptiles. Other extensive alterations
during macroevolution are the production of amphibian legs
and feathered wings on birds. These evolutionary events are
best treated as concepts since they have gray areas. Some
episodes in the history of evolution can appear as medium
size events. The rise of the subclass of marsupials is a large
event, though not as huge as the evolution of amphibians
from fish.

Marsupials appear on the menu as a side dish in an
intelligently designed course from soup to nuts ending with
humans. They surfaced in Europe and North Africa in the
Eocene Epoch and Asia in the Oligocene Epoch. They died
out in those lands and wound up in North America far from
the cradle of human origin. At the end of the Cretaceous
Period, marsupials began moving to contiguous South
America where they settled and thrived. They were lucky

because the potentially dangerous placental mammals were marooned in North America when the two continents separated.

Marsupials had radiated to the large landmass called Gondwana that included Australia and South America. Australia broke away from Gondwana in the Miocene Epoch and was isolated with its marsupial populations. Isolation benefited kangaroos and their kin because the oceans acted as a great moat around their down-under habitat. Across the ocean, the South American marsupials and their North American placental relatives were well adapted to their separate abodes. They remained apart for several epochs and lived like well-tended royalty in their castles.

Nothing lasts forever, and the end of the beginning started in the Pliocene Epoch when South America boomeranged into North America and formed a bridge. The northern mammals crossed into South America and, according to theory, outfoxed the marsupials in the contest of survival of the fittest. Many of the southern marsupials were driven to extinction, excluding the Virginia possum, the émigré and a favorite victim of American motorists to turn into roadkill. Since highway statistics for possum fatalities aren't recorded, it's not known if the motorists kill more possums or other people.

Evidently, the dimwitted marsupials were bested in South America by the brighter mammals, the class crowned with human intellect. Still, human intelligence isn't complimented if it's discovered people kill more possums than themselves. A compliment is surely deserved for paramammals, the marsupial forerunners that replaced the duller dinosaurs in an evolutionary progression. Even so, a progression of small brains enlarging into brains with the intelligence to enable a takeover is questionable. In a takeover, Mayr says, "to actually prove the causal connection in such a sequence is difficult."

The main theory for the demise of dinosaurs is a drastic climate change and not a takeover by smarter reptile-like mammals. After all, the big reptiles and smaller paramammals coexisted in two separate niches in the same environment for 100 million years. Consensus is they lived independently like people in an apartment building until the big eviction extinction. Mayr says the paramammals "were small, insignificant and quite likely nocturnal." The small critters crept out at twilight time to gad about while the diurnal dinosaurs dozed. Apparently, it's easier to live and let live when prey and vegetation are plenteous.

The stationary coexistence of dinosaurs and paramammals isn't puzzling in light of prolonged static periods throughout geologic history. The reef building corals, for instance, remained near genetic equilibrium for millions of years. They changed little because they were unmolested by other organisms and their physical environs were stable. Normalizing selection honed them to their environment as it had stabilized the paramammals and dinosaurs to their world. Natural selection alters organisms when needed, but doesn't act without a reason and they may remain in stasis. The static coexistence of paramammals and dinosaurs may seem a symbiotic mutualism, but it was coincidental.

Currently, there is a static coexistence on the species level between African crocodiles and hippopotamuses. The two human relatives share the same river under an uneasy truce. Although the feared reptiles and big mammals coexist, there is more tension than existed in the noninterference association of the dinosaurs and paramammals. The carnivorous crocs and herbivorous hippos often lay near each other on the riverbanks coming and going at their leisure.

Mother hippos are seen pushing their babies next to recumbent crocs while staring the toothed beasts in the eye.

The mother's odd behavior is hypothesized to emphasize her little ones are not reptilian groceries. The mothers want it known their children are cherished and the croc better keep its snout out of the cookie jar. Mother hippos repeatedly stress their physical dominance over the crocs, which could be necessary because a crocodile has poor associative memories. In other words, the big reptiles simply don't have enough sense to remember who is boss for more than a short while.

Intuitively, the dreadful image of a repulsive croc makes it seem the mother hippo is bluffing. Absolutely not, because a huge hippo can snatch up a half-witted croc in its monstrous mouth and shake the eyeballs out of their sockets. Despite a low or no IQ, the crocs have learned to mind their manners under the vigilant eyes of hippopotamuses. Even so, out of sight out of small mind since opportunistic crocodiles will eat young hippos that stray from the herd. A succulent suckling can be the last meal for a careless croc because death is the penalty for 'hippocide'.

Aaaaaarrrr!

How dare you!

The relationship of crocs and hippos could be a behavioral trait that became heritable by the Baldwin effect. Also, it could be a learned or nongenetic behavior that was passed down from preceding generations. Acquiring a nongenetic trait through group association is correspondent to cultural traditions passed through human generations. Children learn from experiences and by interacting with family and community members. Cultural traits can be subtle, such as facial expressions, or obvious when people talk with their hands or follow a statement with 'eh'.

People recognize people of other cultures by their mannerisms and easier by native costumes. A baby adopted into a new culture will absorb the nongenetic basis of cultural traits. The maturing child absorbs the new culture's characteristics like speaking the language without an accent if the language is different. Observing people acclimating to new cultures is easy, but to determine if the behavior of the crocs and hippos is genetic or nongenetic is difficult. The crocs and hippos would have to be separated for a generation and then reunited.

The behavior of the crocs and hippos has a selective advantage whereas saying 'eh' is an irrelevant custom. Some animals enhance their fitness and survival through cooperative behavior. Mayr says, "any genetic contribution toward high cooperative behavior would be favored by natural selection." Cooperative behavior was observed years ago during studies of societies of human hunter-gatherer tribes and undisturbed chimpanzees in their natural habitats. Inferring closely related genotypes in extinct Homo hominids and Australopithecines indicates those human and chimpanzee relatives lived in cooperative social organizations.

Anthropologist Margaret Power says a cooperative society is egalitarian provided "all members are considered to be of equal intrinsic worth and are entitled to equal access to, and share of, the goods, rights and privileges of their society." Certainly, the tense relationship of the crocs and hippos is cooperative, but it would hardly be called egalitarian. Perhaps after many devoured babies and executed crocs they acquired a nongenetic peace treaty, or Nature endowed them with a genetic proverb to avoid carnage when feasible as in "peace on earth, good will to-."

Credibly, the primate precursor of the hominid ancestral ape selected cooperative behavior. The selection is deduced by the communal lifestyle of Neanderthals and cooperative hunting by Homo erectus. Perhaps the behavioral genes that reinforced cooperation among the ancestral ape's precursor were Baldwinized into the hominid genotype. Through many generations the behavior became heritable and supports the hominid egalitarian society. Mayr says these societies can "be a target of selection," or what works best persists. Further, the early studies of human hunter-gatherers and chimpanzees infer their original society was egalitarian, although both societies have been corrupted to various degrees.

Chimpanzees and humans have thousands of coding genes in their nearly identical genotypes, but 97% of their total genes have no obvious purpose. "There is a widespread belief among Darwinians," Mayr says, "that such apparently unnecessary DNA would have been eliminated long ago by natural selection if it did not have some, as of yet undiscovered, function." Conceivably, the purposeless genes were used by predecessors and are in limbo to be re-employed when needed. Then too, they could have operated in precursors and remain like useless vestiges in descendants.

There are new genes of recent vintage and old genes that trace back to bacterial genes. Genes can act independently or in gene complexes to code for one trait or multiple characteristics. Old human and chimpanzee genes imply their genes accrued between Precambrian bacteria and the emergence of the two primates. This being so suggests the genes associated with primate egalitarian societies were selected with the advent of altruistic behavior. Old genes in Precambrian bacteria that coded for cooperative behavior could be integrated with new genes in animals that adapted egalitarian societies.

"New genes originate by duplication," Mayr says, "with the duplicated gene inserted in tandem in the genome (genotype) next to the sister gene." What's more, the new gene initially has the same function as its sister gene, but may take on another purpose. New genetic directives are acquired by duplicating groups of genes and chromosomes producing different chromosome numbers. Humans have 46 chromosomes and chimpanzees and gorillas have 48 chromosomes. Deductively, the ancestral ape had 48 chromosomes.

"Chromosome 1 in Man may be derived from the centric fusion of chromosomes 13 and 15 of a karyotype like that of the chimpanzee," Chiarelli says. "The correspondence between these chromosomes is evident. A centric fusion seems therefore to be responsible for the reduction in chromosome number in Man from 48 to 46, starting with a common ancestor with the apes." A mutation that fused two chromosomes into one in the ancestral apes might have moved them to the ground permanently or part time. Further, if Australopiths had 48 chromosomes, a reduction to 46 could account for the Homo lineage.

The time expended in reducing two chromosomes into one in the human precursor is not known, and not knowing for sure is typical of evolutionary events. An abrupt evolutionary event is a spontaneous mutation called a saltation, which a theory poses produces a new kind of individual that becomes the progenitor of a new species. Saltationism mimics punctuated equilibrium and contrasts to the theory of gradualism that holds evolution occurs slowly. Many evolutionary scientists dismiss saltations occurring during evolution since species have been strongly honed in their niches by normalizing selection and are reluctant to saltate.

Saltations justify gaps in the fossil record, but gradualists think gaps represent missing or destroyed fossils. Fossils between the early Australopithecines and ancestral ape are scrimpy and give the impression some Pliocene apes were 'saltated' out of the trees. However, it's hypothesized the environment wasn't conducive to fossilization or only a few hominids lived at that time. Certainly, it's possible a saltation by mutation is the reason for a quadrupled ape becoming a bipedal ape. However, it's curious how the phenotype of an Australopith evolved quickly into the phenotype of a Homo sapiens. The Homo hominids present many curiosities.

An uncurious possibility for altering the ancestral ape's phenotype is the failure of chromosomes to disjoin in a genetic 'accident'. The 'accident' is termed nondisjunction and it occurs in the reproductive organs during meiosis when gametes are assembled. Perhaps reducing the chromosome number by two through nondisjuction induced an endurable disorder that natural selection adjusted for bipedal walking. Provided the hypothetical mutation was severe, Nature still amended the afflicted apes as best she could on the African terrain.

Other chromosomal mutations that affect phenotypes are: deletions, when a piece of a chromosome breaks off; inversions, when a broken piece reattaches itself in reverse order; and translocations, when a broken piece attaches to another chromosome. The effects of mutated genes and chromosomes vary from mild to lethal. Skeletal dysplasia, for example, is a genetic condition caused by a sporadic mutation that can be detrimental to skeletal growth and development. Dysplasia produces symptoms from slight shortness to dwarfism or severe impairment.

Chromosome inversion

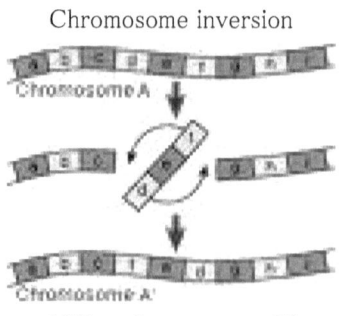

Afflicted some apes??

Dysplasia could have smote the legs of some arboreal apes and induced them to leave the trees. Apes with gimp legs could still climb and descend the trees with their upper body strength. What's more, if they stayed on the ground they may have separated permanently from their unimpaired family and friends. The number of apes and the severity of the condition would determine the time it took them to separate entirely from the parent population. The probability to evolve into a new species was increased through geographic speciation had dysplasia or a similar misfortune brought them to the ground.

Another nominee for bringing the apes down is Marfan syndrome, a heritable condition of humans and nonhuman primates. Marfan is caused by a sporadic mutation in a single gene on chromosome 15. Purportedly, chromosomes 15 and 13 combined to reduce the chromosome number from 48 to 46 starting with the common ancestor with the ape. Some symptoms of Marfan syndrome are an arch in the roof of the mouth that crowds the teeth and spinal curvature that affects leg bones. Conceivably, the condition warped the ape's legs and hastened them into a different adaptive zone.

Whatever condition moved them to the ground, the odds aren't bad it was a reproductive 'accident' or mutagen such as x-rays. The particles comprising x-rays can penetrate reproductive organs and alter genes or chromosomes. Radiation can originate from earth fissures produced by earthquakes or volcanoes. Moreover, fallen radioactive meteorites, comets or rubble from exploding stars emit radiation from their decaying nuclear matter. Betrell says, "a small nuclear reaction occurred spontaneously," in South Africa millions of years ago. Sizable nuclear reactions from atomic bombs dropped on two Japanese cities in World War II resulted in grotesque human mutations.

A fallen comet explodes on impact

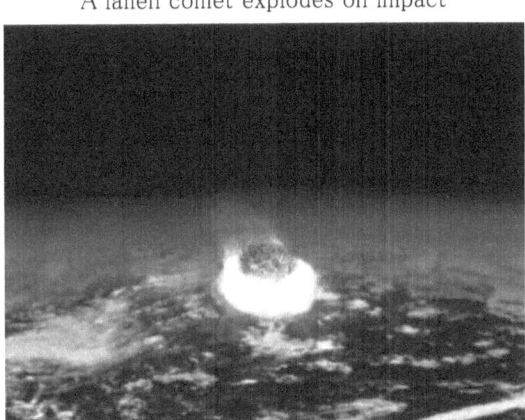

The Radioactive Genesis of Humanity?

The ancestral apes with mutated bodies relocating to new adaptive zones would hurry Nature to act. Adaptations needed to adjust arboreal apes to terrestrial life would come from selecting the best phenotypes with their genes. Judging by the number of knee operations suffered by the ape's

˘current progeny, the corrections needed a better quality of genes. Johanson says a person's knee joint "is a complex mechanism, harder to duplicate and replace than hips or any other human joint." The similarity between human knees and fossils of ancient hominids suggests knee problems have been around for a long time.

Tribulations galore befell people because their ancestors left the trees and adapted bipedal walking. Johanson says the bipedal gait is "one of the oddest behaviors found in nature." The stand-up apes turned into Australopiths and five million years later Homo sapiens evolved retaining numerous vestiges of their ancestors. "One only needs to think of the many weaknesses in humans," Mayr says, "that are remnants of our quadrupedal and more vegetarian past, for instance, the facial sinuses, the structure of the lower vertebral column, and the caecal appendix." Indeed, Nature left people with lots of undesirable souvenirs.

Another memento is the hominid foot that morphed from the ape's grasping foot by the four toes aligning with the big toe. When the big toe curves over the other toes it causes hammertoe, which often requires surgery. The structure of human foot joints does not permit the internal mobility of ape feet. Pilbeam says the human "femur (thighbone) is spared the lower leg weight, but the foot bears every pound." Problems from undue pressure or weight include flat feet, fallen arches, sprained ankles and crumbled spines. Perhaps intelligent design knows why bad backs immobilize their miserable victims.

Ape and human foot

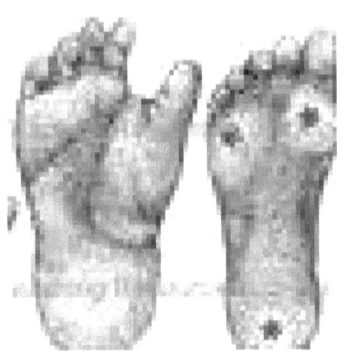

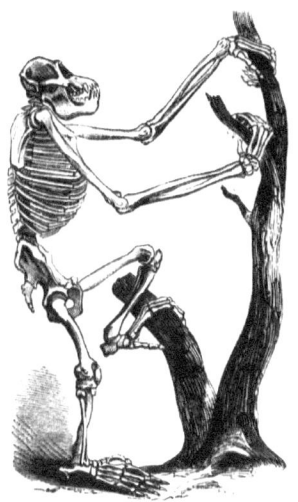

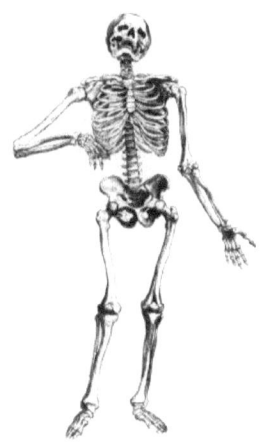

Chimpanzee
Skeleton

Human
Skeleton

CHAPTER 9

The physical woes that plague people and not other primates presumably seeded in the Homo linage after the Australopiths evolved. The Australopiths physical troubles are speculative and assumed small compared to modern humans. Since Australopiths had small brains, it has been suggested that people's problems began with Homo habilis brain expansion that aided survival and reproduction. Yet, the chimp size Australopith brain was ample longer than the Homo lineage has existed. Mayr says big brains don't have a reproductive advantage in a complex society and there is no "trend toward a steady brain increase in the hominid lineage."

Bloated brains could result from regulatory genes that control the pace of fetal development. "A mutated regulatory gene," Mayr says, "may cause a drastic change of the phenotype." However, an extreme change is limited and would not turn an otter into a whale or a chimp into a person, at least in one event. Gould says, "the primary genetic difference between humans and chimps lies in this all-important regulatory system." The genotypes of the two primates differ by a tiny percentage of regulatory genes.

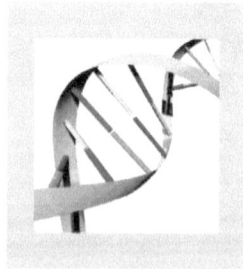

Mutated sequence of regulatory gene

Comparison of Human and Chimpanzee
Chromosomes

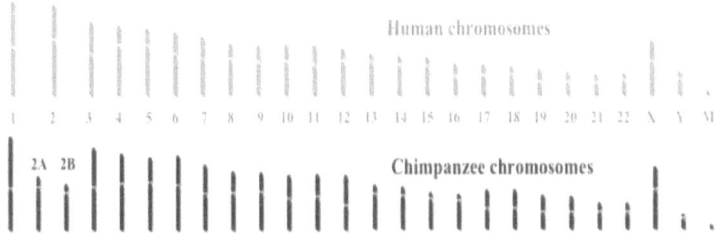

Peas in a Pod?

Regulatory genes apparently account for the odd human body and its juvenile endocrine system that Dutch anatomist Louis Bolk addresses in his fetalization theory. Bolk says an alteration of hormonal balance delayed embryonic maturation and the human body "is a primate fetus." Bolk's theory is based on features people share with the juvenile stages of mammals. Immature traits are the effects of neoteny, or juvenile characteristics retained in adults. Neoteny is adaptive in certain animals such as the water doll, a salamander that remains and thrives in its larval stage like the juvenile Peter Pan who never grew up.

Some juvenile characteristics of humans are: at birth the digits and ends of large bones are cartilaginous, but ossified in other mammals. The hole in the skull, termed the foramen magnum in most mammal embryos points downward and moves backward during gestation, but in humans the opening stays put. The sexual canal of mammalian embryos is ventral and rotates rearward during gestation to facilitate copulation in adults. Conversely, the canal in women remains fixed making it unwieldy for sexual intercourse. Evidently, after nine months of pregnancy, the unfinished human fetus is like a turkey taken out of the oven half-cooked.

"Human babies are born as embryos," Gould says, "and embryos they remain for about the first nine months of life."

The unfinished babies purportedly exit early because the birth canal didn't expand enough during evolution for the big head. "Thus much of the brain had to be shifted to the postnatal period," Mayr says. "At birth the human newborn is essentially 17 months premature." Human infants don't have the mobility and independence of newborn chimps until they are 17 months old. Apparently, natural selection or intelligent design settled for an abbreviated pregnancy that ends with a cordless fetus.

One of various answers to the conundrum of rapid gestation and the jumbo brain is they pertain to social adaptations. Supposedly, delayed maturity was selected to give the brain time to absorb learning and evaluate experiences. "Our advantage lies in our brain, with its remarkable capacity for learning by experiences," Gould says. "Our children are tied to their parents, thus increasing their own time of learning and strengthening family ties as well." That being the case, it's regrettable for the countless women who suffered or died during childbirth that a smaller brain couldn't orient their children to society.

Another supposition for babies exiting early is they need to leave the dark womb for the rich environment of sights, smells, sounds and touches as if they gestated in the twilight zone. While growing in the dark womb, cells that haven't matured into glands secrete hormones that control fetal growth and development. The fetus is dependent on hormones such as thyroxin, and a deficiency of the hormone can retard growth and cause dwarfism. Other hormonal deficits produce physical weakness, infantile sexuality, menopause and multiple conditions absent in other primates. Plausibly, mutated regulatory genes are responsible for the juvenile endocrine system.

The deficiencies emphasize human offbeat features, which also troubled preceding Homo hominids. Mayr says the phenotype of any organism is "likely that of the immediately preceding generations." Thus, people would likely resemble Erectus or Ergaster, which many scientists consider the same species. Still, human DNA suggests people should look like the ancestral ape. Primates "are a little different from the general mammalian pattern, because the childhood period is extended," Leakey says. "But there is no adolescent growth spurt in nonhuman primates. That's a human characteristic, and any parent whose offspring have gone through this stage knows how dramatic it is.

"One minute the child is just that-a child- and the next, he or she is an adult, once the gangling awkwardness is over. If you happen to be a chimp parent, you won't see this overnight transformation in your offspring. Instead, you would see a more steady transition. The extremely extended childhood in humans is the result of a much reduced rate of growth during that period." Extended childhood in a nonadaptive society often includes emotional dependence that never ends. Further, extended physical childhood dependence isn't limited to Homo sapiens according to calculations derived from fossils.

A trained scientist knowing the brain weight of a species "can, with the appropriate mathematics, calculate each of its life history variables," Leakey says. Some variables are body weight, sexual maturity, gestation period, age of weaning, longevity and more. Fossils of a 1.7 million year-old Homo erectus show an assessed brain size of the Erectus newborn of 275 cc. The figures denote that Erectus had fast fetal growth rates that "continued after birth producing increased dependence and prolonged childhood." So far, it's undetermined if any Erectus adults remained psychologically dependent on their parents until they were old hominids.

Comparing the Erectus 10 cc birth canal to the 12.5 cc human canal indicates the fetus also needed to leave the twilight womb a bit early for the rich environment of the outer limits. Homo erectus was similar to Homo sapiens for sure, but Erectus was physically stronger than Sapiens, its direct descendant or nephews and nieces. Leakey says a "professional wrestler would have been a poor match for the average Homo erectus." The weaker human with the large brain suggests bodies weakened as brains expanded, or brawn was forfeited for brain.

Provided bigger hominid brains have higher IQs denotes Neanderthals were the smartest in the Homo lineage since they had the largest brains. Neanderthals are often portrayed as stoop-shouldered stumblebums with low IQs. Ironically, it's curious they had brains larger than the species self-proclaimed to complement the cosmos. The irony is, if bigger brains mean weaker bodies, the skeletal structure of Neanderthals shows they were stronger than Homo sapiens. Thus, the starred brain of the cosmos complement might not be the brightest that ever glowed in the sky.

Whether produced by random selection, or intentional design, human brains and babies are proportionally larger than chimp brains and babies. Scientists assign the average brain size of a mammal species as the brain size of that species. Brain sizes within a species vary because of individual differences, and high IQs aren't positively correlated with bigger brains. Nobel Prize recipient Anatole France, for example, had a 1,000 cc brain volume whereas the brain of erstwhile Lord Protector of England, Oliver Cromwell, was 2,000 cc. The two men were intelligent, but one had a brain twice as big as the other.

Human brains average 1,350 cc in volume and chimps average 395 cc. A liter is 1000 cc or about a quart so a human brain is 350 cc more than a liter and a chimp brain is right at a half liter, or pint. A chimp brain doubled in volume is 790 cc and would still be 560 cc smaller than the human brain. The average size chimpanzee newborn is half of a human newborn, so a chimp baby would slide easily through the human birth canal. Doubled in size its ease of exit would depend on its head size.

Human brain 1,350 cc.

Two-thirds tumor?

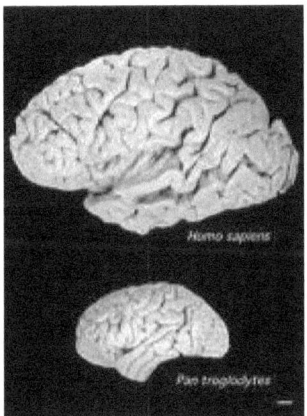

Chimpanzee brain 395 cc

A human baby's 385 cc brain is nearly the size of an adult chimp brain. The human baby has a bulbous head that passes through a birth canal suited better for an ape fetus. A baby chimp brain is 200 cc, which is about half of a 395 cc adult chimp brain. The 1,350 cc human brain is slower than a chimp brain in attaining its adult size. Chimp brains are 70% their final size early in their first year, but even as fast as human brains grow, 70% of their full growth isn't reached until the third year.

The human brain continues to grow at "rapid, fetal rates after birth," Gould says, and is almost adult size when the body is 40 percent fully grown. Ironically, the sluggishly growing human body lags behind the rapidly growing brain, a phenomenon out of sync with the other great apes. The idea that people's retarded growth is needed for more cultural learning is curious. The idea is questioned by the fact most of their existence was in simple hunter-gatherer tribes where cultural learning wasn't needed. Then too, and if it is noteworthy, Homo sapiens suddenly evolved with oversized brains.

The brains of the Homo hominids didn't enlarge by simply blowing up the ancestral ape's brain like a round balloon. On the contrary, during the evolution of the Homo lineage irregular growth resulted in lopsided brains, most pronounced in the humans. The brain's prefrontal cortex determines the intelligence of a primate. Johanson says the "average primate with a brain blown up to the size of ours would still have a prefrontal cortex that is more than 200 percent smaller than that of the human brain." Compared to people, the average primate with the blown-up brain would be intellectually challenged.

Provided the bigger brain was selected because it was needed would explain superior human intelligence, but not its lopsidedness. Fossil skulls indicate Homo hominids brains began expanding 2.5 million years ago. The ape brain sits flat on the basicranium or skull floor, but is arched in humans. Fossil skulls of Homo erectus infer the basicranium is arched in all Homo hominids to a degree. Leakey says the skull floor of an early Homo erectus showed a rise, but in Homo habilis a rise can't be seen since no Habilis skull "discovered so far has an intact basicranium."

Leakey thinks an intact skull of the earliest Homo hominid would disclose the start of a basicranial arch. The enlarging brains supposedly accounted for increasing IQs and intensified birthing difficulties in each new Homo species. Thus, Homo erectus was smarter than Homo habilis, but had more birthing troubles than its precursor. Furthermore, Erectus wasn't as smart as Sapiens, but had less birthing troubles than Sapiens, the last Homo species to evolve. One mystical notion attributes birthing sorrows to taking the advice of a deceitful reptile in the middle of the Holocene Epoch.

Albeit no intact basicranium exists for Habilis, their restored skulls show their brain size was between Australopiths and Homo erectus. The larger brain of Habilis indicates brain expansion and increasing intelligence along with its concomitant tribulations started with Habilis. Conversely, Mayr thinks enlarging brains began with Australopiths since they needed more intelligence to survive when bushes replaced trees on the savanna. "The Australopithecines could no longer escape carnivores by climbing trees and had to depend on their ingenuity. Thus a powerful selection pressure for an increase in brain size developed."

Another brain that wasn't assembled by intelligent design is the chimpanzee brain. Psychologist Wolfgang Kohler says chimps show "intelligent behavior of the general kind familiar in human beings-a type of behavior which counts as specifically human." Louis Leakey, late patriarch of the Leakey family, says chimps haven't changed much since they evolved and were "almost fully differentiated" in the Miocene Epoch. Chimpanzees are termed the common chimp and were living in cooperative societies when bipedal apes emerged. Around the time Homo erectus evolved, a chimpanzee subspecies, the Bonobo appeared on a bank of the Congo River.

Theory holds bonobos speciated rapidly from a founder population isolated by the river. Mayr says when a species is "dissected by geographical and ecological barriers and there is very restricted gene flow in this species, speciation will be rapid and frequent." The slight DNA difference between chimpanzees and bonobos denotes there was sufficient gene flow among chimp populations to limit chimpanzee subspecies. Chimpanzee and human DNA is closer than the DNA difference separating a species from its sibling or subspecies. The DNA similarity convinced some scientists that people should be in a genus with bonobos and chimps.

People and chimps have a potential ancestor in 4.4 million year-old Ardipithecus Ramidus. Paleoanthropologist Tim White discovered female fossils of the primordial primate that was named Ardi. She was 1.2 million years older than Lucy or 60,000 generations at 20 years per generation, but inexplicably the ancient hominid wasn't as apish as the Australopiths. Ardi was assigned another genus because of primitive features such as dentition and jaws smaller than ape jaws. The incisors are smaller than chimpanzee incisors, but larger than Australopith incisors. The lower molars compare to hominid molars since they are broader than an ape's molars.

Ardipithecus Ramidus

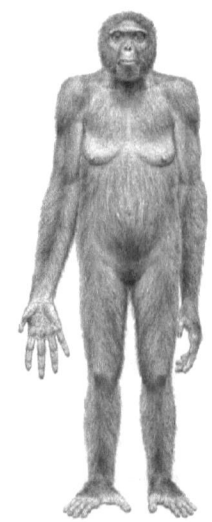

Granny or Auntie?

Ramidus was either an early intermediate hominid or a dead end that faded out without diverging into more species. Ramidus was not fully bipedal, but was adapting upright walking. Ramidus conceivably descended from stooped apes that left the trees because of a malformed skeleton and their foramen magnum wasn't positioned for bipedal walking. Through time mosaic selections situated the hole for an upright stance while contracting the teeth and jaws to oblige the creeping orifice. Ardi shows there was diversity among evolving hominids, and if they were preplanned to be people, they took a roundabout way of getting there.

Ramidus could be the direct precursor of Australopithecus anamensis, a fully bipedal 4.2 million year old hominid discovered by paleoanthropologist Meave Leakey. Provided anamensis is the direct descendant of Ramidus, it follows that Ramidus initiated bipedalism on which Anamensis improved. Some scientists think Anamensis and Ramidus were two of several species or varieties of upright apes that co-existed before the Australopiths evolved. Others think Ramidus was an early offshoot of the ancestral ape and Anamensis gave rise to the Homo lineage.

Scientists who think Anamensis was the forerunner of the Homo hominids are shooting Australopithecus garhi out of the ancestral saddle by nearly two million years. Provided Anamensis was the forerunner means the Homo lineage diverged from the first bipedal ape with the Australopiths or shortly afterwards and muddles the timeline of initial brain expansion. Regardless of what happened, the ancestral apes that left their shaded arbors to become bipedal needed bodily modifications for their brave new world. Perhaps intelligent design partially prepared the Pliocene pioneers for terrestrial life before they left the trees.

Reconstructed fossil skulls of Australopithecus anamensis

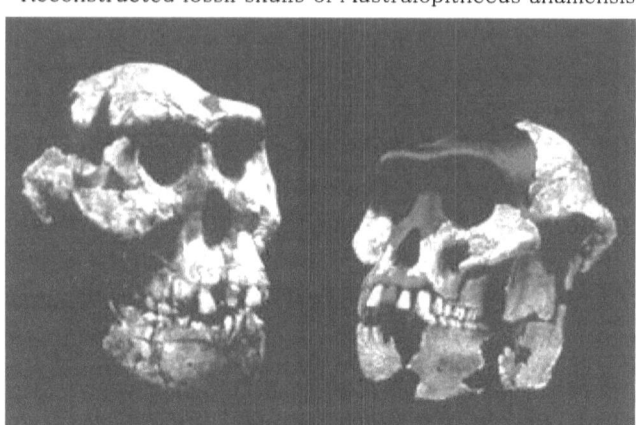

Since African apes walk on their knuckles, it's inferred ancestral apes were knuckle walkers. A practical plan would be to trade knuckle walking for bipedalism while the apes lived in the trees. After all, the robust salamanders presumably began turning into amphibians in the water. Later, the apes, like the salamanders, could make more changes in their new adaptive zones. Darwinians believe nothing was anticipated and Nature produced amphibians and hominids by choosing favorable, but random traits. Whatever sent the Pliocene primates down to earth, theory holds they became bipedal apes, and later Ardipithecenes, Australopithecines and Homo hominids.

Provided the Homo hominids emerged concurrently with the grounded Pliocene primates, or soon thereafter, means undiscovered fossils tell no tales. Hominid fossils are scarce since scavengers devoured their dead bodies or the earth's shifting crust pulverized their bones leaving bits and teeth. Fortunately, Lucy and the Turkana boy established an ancestral timeline. However, a timeline and origin for all hominids such as Homo rudolfensis isn't known. Homo rudolfensis doesn't seem to descend from "any known species of Australopithecus in eastern or southern Africa," Mayr says. "Rather, it seems to have invaded eastern Africa from some where else in Africa."

Rudolfensis had a 775 cc brain and was discovered by Richard Leakey near Lake Turkana. Sexual dimorphism or size difference between males and females was slightly more than in people. Males were 20–30% larger than females. Homo habilis, reputed precursor of the Homo lineage, had a 625 cc brain and was discovered by Louis and Mary Leakey. Two opinions are: Habilis gave rise to Rudolfensis or Habilis forked from the common ancestor and gave rise to the 1,000 cc brain Erectus followed by Homo heidelbergensis with a 1200 cc brain. They are exhibited on the Smithsonian website.

Homo heidelbergensis
Circa 500,000 BCE

Getting smarter all the time

Heidelbergensis evolved 600,000 years ago in Africa, and 300,000 years later gave rise to Neanderthals, and afterward Homo sapiens evolved. Homo brains enlarged to 1,650 cc in Neanderthals, and if intelligent design enlarged brains to raise IQs, people got the short end of the stick with their

1,350 cc brain. Brain expansion, Gould says, "has often been called the most rapid and most important event in human evolution." He didn't say it was a great success or the best thing that ever happened or why the bigger brain of Neanderthals didn't save them from extinction.

One million years ago and 700,000 years before Neanderthals evolved, the Robust species of Australopithecines was the last Australopith to bite the extinctive dust. Their big jaws and molars selected for tough vegetation ceased chomping forever. Analogous to jaws were the beaks of Darwin's mockingbirds that Nature modified to facilitate feeding. Darwin observed three mockingbird species on three Galapagos Islands separated 600 miles by water from South America. He knew only one species lived in South America and reasoned it colonized the islands and gave rise to the others by branching descent from a common ancestor.

"When certain organisms share joint characteristics," Mayr says, and in spite of other differences "it is due to the fact that they had descended from the same common ancestor. Their similarities were due to the heritage they had received from this ancestor, and the differences had been acquired since the ancestral lines had split." Hence, the Galapagos birds adapted varied beaks while diverging from the mockingbird, their common ancestor. Likewise, hominids adapted superficial traits after diverging from their ancestor. Hominid and bird genotypes are built upon genotypes of their precursors, apes and dinosaurs, respectively.

Common ancestor

Similar bird species descended from a
common ancestor.

Different beaks are environmental adaptions for feeding

Another environmental adaptation, but more severe was the
walking apparatus of the Pliocene apes that descended the
trees. All severe adaptations, an opinion holds weren't
initiated by natural selection to acclimate organisms to their
environments. Regardless, the Pliocene primates became
essentially bipedal apes labeled Australopiths that retained
their chimpish brains. Australopiths remained basically
standup apes while their Homo descendants or cousins
transformed grotesquely into mortals. Australopiths didn't
change much of "their ancestral chimpanzee characters,"
Mayr says, "such as small size, large sexual dimorphism
males being about 50% larger than females, a small brain, long
arms, and short legs."

Other dimorphic primates similar to Australopiths, but
heavier are gorillas. Provided gorillas needed to adapt
arboreal life, their phenotypes would require little change.
Certainly, they would need lighter bodies to swing on limbs,
especially the massive silverback, the group guardian. The
smaller gorillas would survive better, thus reducing the

average gorilla's size. Reducing creatures is sometimes practiced by Nature such as dwarfing the Hobbits of Flores, Indonesia. The extinct Hobbits were reduced by Nature to help them survive on limited resources. Natural selection has trimmed down other hominids like pygmies in her determination in promote survival and reproduction.

Promoting survival and reproduction in the Homo lineage, natural selection made more adaptions on them than on Australopiths. An adaption, Mayr says, is a "property of an organism, whether a structure, a physiological trait, a behavior, or any other attribute, the possession of which favors the individual in the struggle for existence." A few ecological or environmental adaptions of Homo sapiens are: Asian eyes, sometimes called the Mongolian fold, wiry and straight hair, skin coloration and diverse body structures and sizes. These superficial traits are coded by a minute percentage of coding genes in the human genotype.

Presumably, Nature adapted fewer superficial traits for Australopiths because they remained in stable African environments. When the ancestral apes became bipeds, their legs were altered more than their torsos, which remained apish. Curiously, it's questionable why the alterations were needed in the first place. Apparently, the apes were as mobile as the chimps Goodall first observed through binoculars. She says they "careened down the opposite mountain slope and began feeding in some fig trees." They knuckle walked rapidly or moved "in an upright position." Sometimes they ran across open spaces alone or in groups going to different fruit trees.

Bonobo walking bipedally

in a zoo

Chimps are agile and catch prey animals such as bush pigs on the ground or monkeys leaping through trees. Chimps can run faster than an Olympic sprinter, which is OK, but not amazing since the sprinter couldn't catch a chicken. Intuitively, the Pliocene apes that traveled on the ground could have done as well on their own legs without turning into bipeds. A simple, though unconfirmed explanation for knuckle walking apes becoming bipeds is a mutated regulatory gene. Mayr says a mutated regulatory gene may result in a "drastic change (discontinuity) of the phenotype."

Mutated genes as a source of new genes cause changes in organisms, or else evolution would remain static. Gradualism proposes that organisms evolve as they adjust to gradually changing environments. The most suitable traits and their tagalong genes are selected during changes to assure the continuity of a species. The best traits, for instance, were selected for whales and horses during their fifty million years of evolution. Their intermediate species and last descendants did not change drastically like Homo sapiens changed while evolving from bipedal apes. Severe changes do occur, but gradualism holds they change over extended time.

Gradual changing from quadrupedal apes to bipedal apes is questionable because of a lack of fossils. Punctuated equilibrium appears to be the candidate since Australopiths emerged suddenly, but new fossils could offer another story. What is certain is the apes retained many vestiges of their previous lives, though vestigial genes remaining in descendants weren't selected against. Vestiges may be problematic or without consequence lending credence to the opinion evolution has no purpose. The feeble wings of flightless ostriches, for example, cause no apparent problem whereas eagle wings are superb adaptions for flying.

Albeit ostriches and eagles share the avian genotype, the ostrich garnered proficient walking genes while perchance relegating its vestigial flying genes to a repertoire of noncoding genes. Generally, organisms would have fewer challenges if intelligent design eliminated vestigial impediments. Regardless, most would be unaware of their existence while a smidgen would be conscience of their lives to different degrees. Intelligent design's notion that Nature is purposeful and perfects organisms is disputed and one reason is evolution's many oddities. True, natural selection is deterministic, but its imperfect adaptions and failure to stop extinctions argue against purpose.

CHAPTER 10

Imperfect adaptations are notable by those imposed on the
Homo hominids by gradual selection, punctuated equilibrium
or another mode. Conversely, Australopith adaptions caused
few problems compared to those of their Homo relatives.
Australopiths persisted with few physical changes for nearly
four million years, longer than the Homo lineage has existed.
Clearly, Nature responds to necessity with physical changes
to adjust organisms to new adaptive zones. Still, it's curious
why evolution was rough on the Homo hominids and easy on
Australopiths. Whatever reason, the Australopith's peculiar
gait bequeathed to all hominids suggests that intelligent
design daydreamed and neglected excellence.

Obviously, random or designed selection doesn't perform
ideally, or there wouldn't be knee or back operations, except
for accidents like slipping on a banana peel. What's more,
Neanderthals would still be around unless they were absorbed
by early Homo sapiens. "Some enthusiasts have claimed that
natural selection can do anything," May says. "This is not
true." The diligence of selection doesn't waver, but there
are limits to its effectiveness exemplified by endless
extinctions and the human body. Intelligent design might not
wish to take credit for that body unless most of the effort was
expended on building a big brain.

On the other hand, intelligent design or natural selection
left the Australopith brains the size of a chimp's brain.
Whichever evolutionary architect designed Australopiths,
their few foibles were restricted primarily to their legs. The
prevalent theory poses Australopiths were adapted for
bipedalism by modifying the lower half of an ape's skeleton in
response to changing environs. An opposing theory holds the
modified skeleton was not initiated by Mother Nature, but by

an unidentified mode of evolution. Regardless, mosaic selection was rather complacent with the upper half of the ape's body.

An example is Lucy's rib cage that Leakey says "turned out to be conical in shape, like an apes, not barrel-shaped, as would be seen in humans." Lucy was essentially a bipedal ape, "and her shoulders, trunk and waist also turned out to have a strong apelike aspect to them." Apparently, once the ancestral apes became Australopiths they became relatively static with minor alterations. Other Australopithecine features Johanson says, are "a short femur, slightly curved finger and foot bones, narrow rather than broad fingertips, a highly mobile wrist, powerful arms, and so on are unlike those found in modern humans."

Lucy's skeleton.
Circa 3.2 million BC

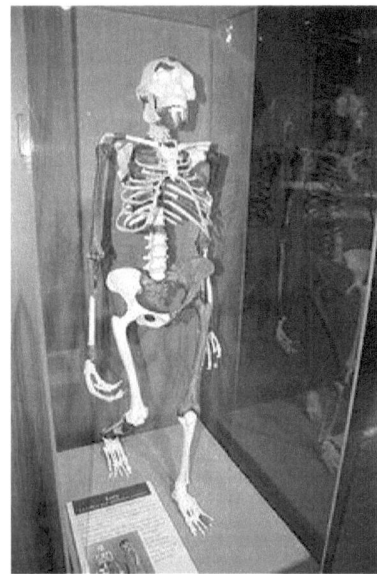

Gone, but not forgotten

The human body compared to an ape's body is flabby and bloated and not just because of unhealthy lifestyles. A chimpanzee the size of a person would weigh much more since chimps are denser than their squashy relatives. One reason for the flaccidity is the ends of people's limbs are packed with shock-absorbing material. Johanson says the material makes the skeleton "bigger and puffier like a sponge tossed into a bucket of water." This physical gap between people and chimps is puzzling since their nearly identical genotypes should code for bodies almost like two peas in a pod.

The pea pod simile isn't apropos for the genotypes of Homo sapiens and Australopiths since Australopith DNA isn't available. However, fossils show there is a physical divide between humans and Australopiths that were basically bipedal apes. Mayr says "except for bipedalism and some tooth characters, the Australopithecines shared almost all their other characters with the chimpanzees. And, what is surely more important, they had none of the most typical Homo characteristics." Deductively, if Australopith and human DNA are ever compared, it will reveal two genotypes almost like two peas in a pod.

Provided the DNA of chimpanzees, Australopiths and the Homo lineage is nearly identical denotes the genetic basis for bipedalism was the main difference between the ancestral ape and its descendants. The ancestral ape's lower bones that were modified for bipedalism have analogous bones in descendent hominids. A fossil of an Australopith knee, for instance, shows the knee was as intricate as the human knee. "Three bones come together here: the femur, the tibia, and the circular kneecap, or patella," Pilbeam says. "In addition a delicate arrangement of injury prone tendons and ligaments surround the bones."

The delicately arranged Australopith knee seems it was potentially as troublesome as the human knee by comparison. However, the Australopith knee could have been hardy while the human knee is vulnerable because of the puffy skeleton. The puffy skeleton could be a result of an inconsistent supply of hormones secreted by the juvenile endocrine system during gestation. Conceivably, the disrupted hormone supply foiled calcium assimilation leading to many human physical frailties, including bad knees. Conversely, the lighter weight Australopiths with a mature endocrine system might have scurried over hills and dales on sturdy legs in sync with durable knees.

The Australopiths scurried intrepidly or precariously with an odd step presently practiced by only one of two hundred primates. Pilbeam says in bipedal walking, "as one leg is lifted off the ground, the pelvis rotates around the leg that remains in contact, thus moving the free limb forward. As the leg is raised, the pelvis tends to collapse but is prevented from doing so by muscular contraction." Moreover, the small gluteal (rump) muscles run from the side of the hipbone to the top of the femur.

The gluteal contraction prevents the pelvis from slumping to the unsupported side when one foot is in contact and it also rotates the pelvis around the fixed limb during walking. The largest rump muscle, or gluteus maximus, acts as a hip extensor, originating on the back of the pelvis and sacrum and running to the back of the femur. It also serves to check forward momentum because the advancing limb acts as a brake on coming into contact with the ground. In living apes these muscles act mainly as abductors to swing the leg outward and aren't major hip extensors.

Somewhere during Homo hominid evolution kinks developed such as weak abdominal muscles. Abdominal and back muscles pull against each other like guy wires to support the backbone. Degenerate and herniated spinal discs are often attributed to weak supportive muscles. Obesity tends to produce excessive strain on the vertebra and pressure on the knees. Many people never have back trouble, which might be accredited to their life style and genetic variation in body structure. "Consider the most successful and least injured running backs in football," Lovejoy says. "They tend to be the players with big trunks and short, stocky thighs."

A price diligent people pay to help prevent back problems is maintaining a proper body weight and doing sit ups to strengthen stomach muscles. People dedicated to an exercise regimen usually don't have bad backs or other debilities resulting from sedentary living. The human body's weakness, like other human anomalies, conceivably results from the juvenile endocrine system secreting an uneven supply of hormones. Since there is no evidence Australopiths retained juvenile characteristics, it implies they were spared most of the physical miseries that haunt their larger brain relatives.

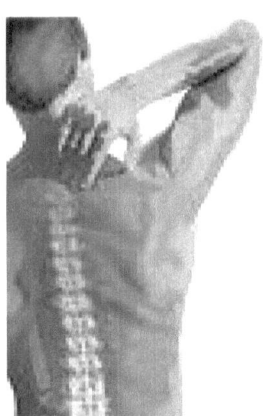

My aching back!

One gigantic grief that historically caused the deaths of untold human mothers and babies are complicated childbirths. Prior to modern medical technology pregnancy was nine anxious months. Two obstetric crises generally remedied by Caesarean section are breeched babies positioned feet first and babies too large to pass through the birth canal. Verily, the multitudes of tragic births in olden times could give the impression intelligent design was disappointed with women and punished them with a shoddy birth canal or undersize pelvis.

Another possibility for obstetric difficulties is mosaic evolution couldn't supply a bigger pelvis for a larger birth canal, or build a larger canal for the pelvis. Whatever caused the birthing quandary visited on the Homo lineage has not been addressed. Plausibly, it was natural selection determined to correct errors with random variation that didn't offer the appropriate genes. Then again, like shattered Humpty Dumpty who all the king's men couldn't restore, it could be there was no solution. "Once a particular body structure has been acquired," Mayr says, "it may not be possible to change it again.

For instance, in terrestrial vertebrates the respiratory tract from the oral cavity to the trachea crosses the digestive tract, which also runs from the oral cavity to the esophagus. This arrangement was adopted in rhipidistian fishes, our aquatic ancestors. Although it poses forever the danger of food entering the trachea, no reconstruction of this inferior pathway occurred in several hundred million years." Sadly, when someone chokes to death on a rhipidistian fishbone it denotes natural selection or intelligent design sometimes can't win for losing.

Another losing arrangement evolution short-changed women with is their birth canal that's inferior to the canal of their chimp cousins. The chimp pelvis is tall and narrow with two femurs that extend straight down and has plenty of room for a birth canal. In contrast, bipedal modification of humans and

Australopiths made their pelvises wider and shorter and angled the femurs in toward the knees. Johanson says when a chimp and human pelvis are shown with a cast of the Australopith Lucy's pelvis in an elementary school "the children have no trouble deciding which two look alike."

Looking alike isn't acting alike because women don't have babies easily like Australopith mothers. Childbirth is often difficult for women and in olden times was called a curse. Justifying the hardship as a penalty for disobedience an ancient scribe wrote "in sorrow thou shalt bring forth children." (Genesis 3:16) Bringing forth children was easier for Australopith mothers thanks to chimp-size babies with small heads. Mayr says the Australopith birth canal "allowed the passage of only a small head, and thus the small brain had to be large enough to serve the newborn and the limited demands of an australopithecine."

Humans, unlike Australopiths and chimps spend 35% of their lives growing following a precarious passage from the womb, but they aren't alone. Fossils of extinct Homo species indicate they had, to some extent, birthing sorrows, retarded maturation and presumably helpless babies. Nevertheless, the blue ribbon for quirks goes to people with the most prolonged stages of infancy, childhood, and adolescence. People, and apparently extinct Homo hominids, are characterized by superior intelligence and physical retardation attributed to big brains. The brains arrive in big heads atop big babies that are squeezed as they exit the womb.

A theory holds brains expanded out of necessity, often called the mother of invention. Responding to necessity and concurring with Darwinism, natural selection chose bipedalism for the ancestral apes leaving the trees. Curiously, no evolutionary mode enlarged brains until necessity beckoned to natural selection for bigger brains. Many scientists say expanding brains generated more intelligence that was necessary for the endangered Homo habilis hominids to survive. Habilis was clever enough to chip hand tools from stone and evidently the first hominid to do so. The tools were found with Habilis fossils among animal remains implying they butchered animal carcasses.

Then too, if the swelling brains were incrementally weakening Homo hominid bodies, tools would compensate for their loss of strength. Another thought is Habilis accidentally used a sharp stone like the accidental discovery of x-rays. As habilis continued using tools, perhaps fashioning some into spears, it's hypothesized natural selection intensified to adapt dexterous hands. Still, nimble hands and other human-like characteristics could be spinoffs of inflating brains. Moreover, inventing tools was perchance an energy saving tactic to make life easier. Making life simpler in modern times and to increase profits is to build a better mousetrap.

Making their lives simpler, laboratory chimps invent solutions to experimental problems bolstering the opinion that enterprise is inherent in chimpanzees and humans. "The human economic system," Franz de Waal says, "with its reciprocal transactions and centralization, is recognizable in the group life of chimpanzees. They exchange social favors rather than gifts or goods, and their support flows to a central individual who uses the prestige derived from it to provide social security." De Waal is referring to captive chimpanzees that do not represent a society of undisturbed chimpanzees in their native habitats.

Chimps in captivity are usually peaceful, thanks to some peacemaking individuals, but tensions in confinement have led to conflicts and deaths. The deaths are thought to be premeditated murder, which seems to mirror the violence of Western society. Conversely, undisturbed chimps in their native environs can avoid aggressive behavior and maintain peace by walking away from stressful situations.

Unfortunately, captive chimps don't have the freedom to come and go like they do in their natural habitats. Furthermore, chimps in their natural habitats make tools from various materials for sundry tasks, remindful of their extinct relative, Homo habilis.

On the other hand, their gorilla cousins aren't known to make tools in their natural environs, but domiciled gorillas make tools for present and anticipated purposes. The late Dian Fossey, who was sponsored by Louis Leakey to study gorillas in Africa, says in her book, *Gorillas in the Mist,* "Since gorillas mainly eat vegetation, food preparation involves manual and oral dexterity, attributes with which gorillas are well endowed. Perhaps for this reason gorillas have not yet been observed fashioning objects within their environment as tools." Obviously, when gorillas need tools they are smart enough to make them.

Chimps use their smarts to crack nuts with stones and fashion stems from grass to 'fish' for termites and honey. They use leaves to sponge water, for napkins and to scoop out the last tidbit of brain from a prey monkey's skull. Goodall cites several instances of chimps foreseeing a future termite meal by selecting "a grass stem for subsequent use as a tool," even when the termite heap is out of sight. Chimps don't need flaked stones to butcher carcasses because they are strong enough to tear prey animals apart with their hands.

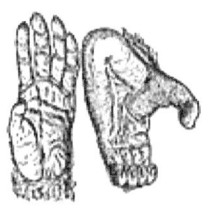

Chimpanzee hands

Making a tool for "fishing" for termites

Laboratory chimps analyze and resolve problems like finding hidden treats or fetching out of reach toys. They figure out how to chip sharp edges on stones and cut strings holding up a suspended box of bananas. Chimps solve spatial problems by stacking boxes they stand on to retrieve food dangling on strings. Moreover, when put in a position of direct competition they quickly learn to deceive. They withhold the whereabouts of hidden and limited treats from other chimps. Truly, in settings akin to directly competitive human societies a chimp doesn't waste a good mind.

"The startling resemblance of basic cognitive mechanisms," Goodall says, showed the uncanny similarity between chimpanzees and humans. Goodall tells of a home raised female that displayed "the various human behaviors she had acquired over the years." She "opened the refrigerator and various cupboards, found bottles and a glass, then poured herself a gin and tonic." She surfed TV from "one channel to another then, as though in disgust, turned it off again," like another primate disgusted with television. Her behavior is remindful of a bored housewife or Mr. Mom with too much time on their hands.

The domiciled female passed more time by flipping through a magazine and finding a picture of a dog she acknowledged with signing "this mine" in ASL or American Sign Language. Hearing impaired people use the silent language that is a commutative method taught to laboratory apes. Goodall says the home-bodied chimp had learned "the use of possessive pronouns." Chimpanzee intelligence raises a question about the survival of early Homo hominids. Their skull floors started arching as their brains purportedly expanded to increase IQs. The high IQs of chimps would seem sufficient for those supposedly threatened hominids.

Gorillas also have high IQs and can 'talk' through signs in ASL and other symbolic communication devised for apes. Gorillas and chimpanzees reveal abstract thinking by inventing signs for objects when the signs don't exist. Young apes learn symbols much easier than adults like human children learn to speak a foreign language faster than grown-ups. Through artificial languages trained apes communicate wishes, ask questions and express love, compassion, generosity, frustration and anger. One female chimp became anxious when she anticipated the reoccurrence of an unpleasant experience, but didn't have the signs to show her concern so she used body language.

"Although chimpanzees have made considerable progress along the road to humanlike love and compassion," Goodall says, they can't say I love you or I feel your pain. A project to teach a home-raised chimp to verbalize failed despite intensive training. The chimp "learned only four words---- which she 'breathed' rather than spoke: papa, mama, cup, and up." Goodall says the vocal structures of chimps makes it impossible for them "to produce the human vowels a, i, and u, and that they lack a certain mobility of the tongue, which humans use to change the shape of their vocal tract."

Chimpanzees and gorillas are smart primates, but when they try to speak they sound like patriots singing their national anthem with a mouth full of marbles. Human speech is enabled by an aggregate of coordinated anatomical features such as the larynx (voice box), pharynx (throat cavity), lungs, ribs, muscles and nerves. The larynx in all primates, except humans, is high in the throat and limits the sounds produced by the pharynx. Human babies are born with a larynx high in the throat so they can simultaneously breathe and swallow, which adults can't do.

The larynx starts moving down the throat after eighteen months, the age babies usually begin to vocalize. Leakey says the larynx reaches "the adult position when the child is about fourteen years old." Still, the descending larynx doesn't solve a big quandary since pre-teens and adults can die by choking. The larynx's position in extinct hominids isn't known since soft parts decay, but knowing its location would help estimate their language capability. Limited language use by Homo erectus was inferred by fossilized basicrania, but nerve orifices in the spine indicate breathing control for speech would be somewhat hampered.

Human speech was favored by natural selection through transferring cultural information from one generation to the next. Mayr says speech "necessitated the origin of language." Conversely, speech isn't needed by the talented and observant chimpanzees to transfer cultural practices such as termite fishing. Goodall says chimpanzees are at the dawn of cultural evolution and the achievement of a gifted chimp "can spread through a group and rapidly become part of its tradition." The continual transfer through generations of skills learned by observation conceivably are totally or partially Baldwinized in the genotype.

Chimps in their native African habitats adapted cooperative social structures while passing on nongenetic customs without a word spoken. Goodall says chimps "have a rich repertoire of sounds, postures, and facial expressions that facilitate the exchange of information among community members." Speech and the peculiar human brain purportedly took root in Homo habilis. Brains swelled from the roundish ape brain to the lopsided human brain that stopped surging and left disproportionate areas like the prefrontal cortex. Some cranial structures outpaced others, which possibly affected the thoracic cavity and repositioned the larynx.

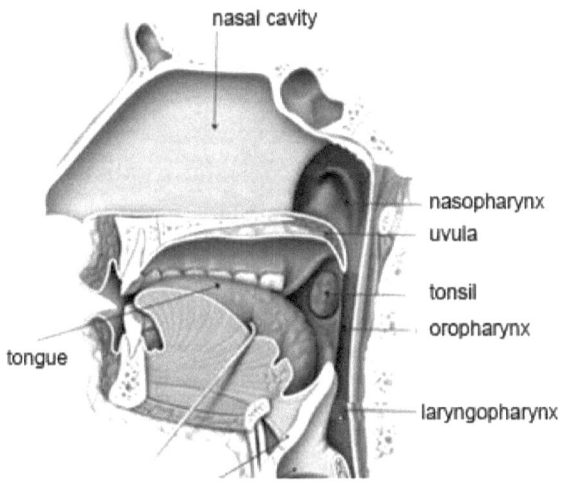

Larynx pushed down by the swelling brain??

Another proposal for descending larynxes was to make loud sounds that exaggerated the body size of a hominid. The sounds alerted a predator to beware of a big animal, although the bark of the bluffing hominid had no bite. Sounds entering the human brain are understood in Wernicke's area, and in Broca's area language elements are sequenced. The connected areas are part of the frontal lobe and one sends and the other receives information. In ape brains the areas are smaller and believed to have other purposes that prevent those intelligent primates from understanding human speech.

The mosaic expansion of Wernicke's and Broca's areas in the Homo lineage presumes natural selection favored the areas for speech. An opposing opinion is the brain ballooned for another reason and speech is a by-product of the bulge that wasn't started by natural selection. The bulge affected parts of the human head such as jaws, maxillary bones and sinuses. Rapid and uneven growth caused the skull floor to arch, which squeezed the jaws and maxillary bones. The disproportionate expansion crowded the teeth and produced impacted molars while distorting the sinus hollows.

Sinuses in apes are thought to have a structural and not a functional use. During human evolution the sinuses were misshaped to a point they often become a medical ailment. Bacteria in sinuses can cause constant irritation and severe infections. Conceivably, the sinus hollows resulted by opposition from other parts. Mayr says the efficiency of selection depends much "on the resistance of other structures and other components of the phenotype." Clearly, there was plenty of cranial resistance to the brain that expanded so fast it doesn't appear to be a product of gradual selection.

The swollen brain's extreme asymmetry might explain behaviors like homosexuality that distinguishes people from chimpanzees. Goodall says she never saw evidence of chimpanzee homosexuality. "A male may mount another in moments of stress or excitement, clasping the other around the waist, and he may even make thrusting movements of the pelvis, but there is no intromission. It is true, also, that a male may try to calm himself or another male by reaching out to touch or pat the other's genitals; while we still have much to learn about this type of behavior, it certainly does not imply homosexuality.

"He only does this in moments of stress, and he will touch or pat a female on her genitals in exactly the same context." A chimp named Figan frequently touched his own scrotum when he suddenly became apprehensive. Other great apes such as gorillas and bonobos also indulge in various pseudo-sexual doings in moments of excitement. Fossey says an estrus female in a gorilla group "prompts a great deal of vicarious sexual activity among other group members such as mountings between individuals of the same sex or between animals of different age groups.

Gay Celebrities
The bloated brain again??

Elton John   Ellen Degeneres

When a male chimp watches a couple copulating he may experience a vicarious sexual arousal. Goodall says the male's "penis becomes erect, and he may then approach and mate with the same female immediately afterward." Clearly, it's share and share alike. The human primate practices sex by proxy through voyeurism and pornography although it's restricted to a mental pleasure with no reproductive advantage. Inside the human brain the visual images are magnified into larger sexual excitements. However, vicarious sex enjoyed through pornography is not limited to men and women, and in modern society, to teenagers.

Goodall says a domiciled female was given the graphic magazine *Playgirl*. As she stared at the penis of each naked male she moaned uh, uh. "When she came to the centerfold she spread it on the floor, positioned herself over it and rubbed her vulva back and forth on the penis for about twenty seconds." Plausibly, a sexual image in the visual cortex of ape brains is a reproductive strategy ingrained in the genotype. Certainly, the voyeuristic male increased his chances of representation in the next generation by mating after watching a couple copulating in a live peep show.

Edition of Playgirl

Sex Sells

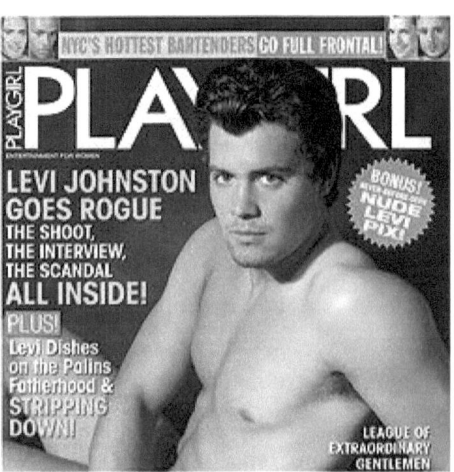

Transpecies stimulant

CHAPTER 11

Celebrated sex shows are put on by bonobos, descendants and subspecies of common chimps. Bonobo adults, adolescents and children wildly embrace, French kiss, rub their genitals together and thrust their hips in the undulating movements of pseudo-intercourse. Group sex parties in the bipedal world featuring swinging couples resemble bonobo sex shindigs, but bonobos have other objectives. "Perhaps we should not call all of this 'sex', because people tend to think of it as a self-contained behavioral category aimed at an orgasmic climax," De Waal says. "I have never seen ejaculations using sex between males nor attempts at anal penetration."

The physical intimacies of bonobos masquerade as sexual foreplay or copulation, but De Waal says the real reason is to "relieve tensions." Apes in their natural habitats experience some tension simply by the requirements of daily living. Their adaptations of wild sex and fission dissipate the tension, which isn't possible in zoos. A zoo is analogous to a penitentiary where prisoners can't walk away from trouble. When chimps fight they usually reconcile later with friendly contact. De Waal says physical contacts after conflicts are "much more intense than contacts in other situations, kissing being the most characteristic behavioral feature."

Reconciliation

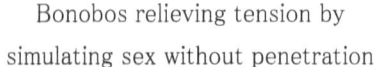
Bonobos relieving tension by
simulating sex without penetration

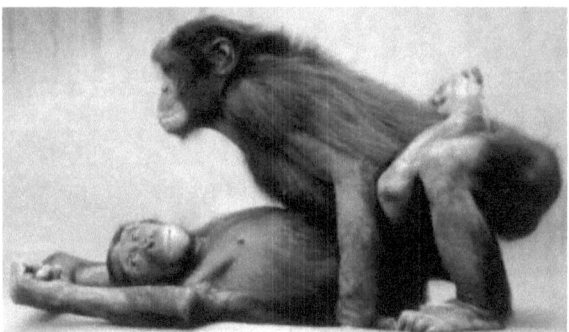

Tensions heightened by fear and apprehension permeate the lives of chimpanzees in their present day habitats. The peaceful habitats were corrupted millennia ago when people started hunting chimps for food. Fear of humans increased when profiteers began capturing them to sell to zoos or as pets or medical research. Goodall says when she first approached chimps in 1960 "they had almost always ran away in fear. To prove myself utter harmless, I feigned disinterest and pretended to chew up leaves and stems." The chimps gradually became habituated, or accustomed "to the white-skinned upright ape," and ultimately tried to intimate her.

Upright humans relieve tensions, reconcile troubled relationships and strengthen bonds by hugging, kissing and copulating. Couples have intercourse not only for pleasure and to reproduce, but for reciprocal assurance in the uneasiness of modern society. Same sex couples practice homosexuality for the same reasons, but obviously not for procreation. Homosexuality is practiced exclusively among human primates and it's hypothesized homosexuality was adapted by primeval hunter-gatherer tribes. The hypothesis poses the sterile practice was chosen so homosexuals could assist with child rearing. Natural selection would have favored a homosexual gene, which perhaps was Baldwinized into the genotype.

"We have identified no homosexuality gene and we know nothing relevant to this hypothesis about the social organization of our ancestors," Gould says. "It need not be vindicated-and must not be condemned-by genetic speculation." Likewise, the unorthodox practice of bisexuality can neither be vindicated nor condemned by genetic speculation. "The prevailing unorthodoxy in the ancient Greco-Roman culture was bisexuality," Scroggs says. Homosexuality and bisexuality both "emerged out of the social matrix of the day. In some quarters they were extolled, in all quarters condoned.

"They were practiced by a large number of people in part because it was socially acceptable. Their culture can fairly be said to be bisexual, since many adults would be married and carry on sexual relationships with both sexes." Greco-Roman sexual custom suggests human orientation toward male or female sexuality is altered when sexual inhibitions don't exist or are relaxed, or before cementing orientation. The extent of social and heritable influence on unorthodox sex isn't established and is a continuing debate of nurture versus nature. The influence of past and present theocracies has a negative effect on unorthodox sexual practices.

"If a man lies with a man as with a woman, both shall be put to death for their abominable deed." (Leviticus 20:13.) Rigid control by pragmatic rulers favored a higher birth rate that supplied more workers for lower social levels. Laws forbidding extramarital relations and nonreproductive sex precluded tolerance of gay blades. Since rank has privilege, the switch-hitting King David exempted himself from the ancient laws. David told Jonathan "thy love to me was wonderful, passing the love of a woman." (2 Samuel 1:26) During David's

heterosexual adultery with Bathsheba, the pragmatic king
engineered her solider husband's death.

David and Bathsheba

David and Jonathan

Homosexuality and bisexuality, Scoggs says are masked in a
closet mentality that "is still very much a part of the ethos
today." Sodomy and bestiality deviate from the norm and
were prohibited by the ancient Hebrew culture, still the root
of zealous fundamentalism. "Neither shalt thou lie with any
beast to defile thyself therewith: neither shall any woman
stand before a beast to lie down thereto: it is confusion."
(Leviticus 18:23) Many deviations the Old Testament
tabooed are accepted in modern societies, but some are
illegal. Many psychologists attribute the loathsome practice
of necrophilia to insanity.

There is no consensus on the psychological cause of
pedophilia, a widespread criminal offense. Baffling Oedipus
and Electra complexes plus other sexual obsessions are
treated by specialists in abnormal psychology with dubious
results. Clues to the riddle of offbeat eroticism might be

found with fossils of Homo habilis. The flaked stones discovered with their fossils were chipped mostly by right hands. Since chimpanzees favor neither hand, it's inferred right- handedness began concurrently with brain expansion in Habilis. Provided the warping brains induced right-handedness in Homo hominids, the warps might be responsible for sexual quirks people practice.

Quirks could be inherent in 'civilized' people being oriented sexually through obligate learning. Humans are born at a primitive developmental level, and if they are born sexually neutral then sexual preference is determined during maturation. The path a person takes, straight or circuitous or multidirectional, is affected by family, society and obscure variables. Homosexuality and other oddities denote arbitrary paths while bisexuality suggests sexual neutrality was unaffected. Perhaps Voltaire's opinion is true that the happiest people live in primitive tribes oblivious of modern social controls.

A physical explanation for the source of deviance is defective nerve connections between the cerebral hemispheres of the lopsided brain. Provided the cerebral nerves were misaligned by the swelling brain is analogous to rigging a machine with faulty wiring causing a short circuit. A sexual short circuit fusing paternal instinct and social immaturity might be responsible for the puzzling perversion of pedophilia. Indeed, highly speculative, but nothing has explained the profound sexual disparity between humans and chimpanzees with 98.8% of the same genes.

Moreover, sexual disparity between the two primates conceivably resulted from brain enlargement being too fast for natural selection to build cerebral hemispheres with precise circuits. Brain expansion "has often been called the most rapid and most important event in human evolution," Gould says. He didn't say it was a job well done or the best thing that ever happened. Brain expansion began in Homo hominids and stopped pressing inside their skulls with the speciation of Homo sapiens, leaving them with mystifying characteristics. On the contrary, Australopith brains remained the size of chimp brains and void of mystery.

The mystery of the human brain is intricate and possibly involves constraints that hinder a part of the phenotype from acquiring optimal efficiency. The constraints on the expanding hominid brains first had to be altered or removed for the brains to swell to their perplexing dimensions. One deduced constraint is the vascular, or blood network that cooled the brains of the ancestral apes that left the trees. Patterns of the blood networks of extinct hominids can be examined since cranial vessels are imprinted on the inside surface of fossil skulls.

Mammal brains are heat sensitive and the efficiency of their cooling system determines how they adapt to hot niches. The human brain, for example, malfunctions when its temperature increases a few degrees. Elevated brain temperature can cause convulsions followed by death. When some descendants of the ancestral apes moved to hot African plains their cranial circulation was minimal for cooling. Still, their brains couldn't enlarge more than their cooling system allowed, like a car radiator won't cool a big truck. Anthropologist Dean Falk's radiator hypothesis poses the network of cranial veins helped regulate brain temperature as the network acclimated to warmer environs.

Mayr says "every colonizing species has to become adapted to the prospective niches it encounters in a newly occupied area." Hence, the blood network cooling the brains of bipedal apes colonizing new and sunny environs needed reconditioning. Nature reconditioned those intricate vascular routes that warm or cool the brain. The routes originate from the carotid arteries and permeate the cranium, which bolsters the individual's fitness. Blood cooled by sweat evaporation at the body's surface circulates through the cranium. The cooler blood absorbs heat in the cranium and exits as warmer venous blood, circulating like water in a radiator.

Another route takes blood cooled at the head's surface and delivers it into the braincase in case of overheating. Conversely, in excessive cooling like hypothermia in icy waters the chilled blood leaves the braincase through a route that would be unnecessary in equatorial Africa. The network that drains "blood from the Homo brain is conducive to efficient cooling," Leakey says, "while in Australopithecines it is much less so." The less efficient network implies Australopiths didn't move into hot areas. Still, if some did move to hot dwellings, their network might have cooled their brains, but couldn't initiate brain expansion.

The different blood networks of Australopiths and Homo hominids support a minor theory that the first bipedal apes branched into two lines. Later the lines diverged into two genera making the Homo hominids Australopith cousins and not their descendants. The larger brain of the Homo genus is an adaption that, by one definition, is a property of an organism that adds to its fitness. There are two ways to prove the adaptedness of a feature like the big brain, Mayr says. "One is to try to show that the occurrence of the feature cannot possibly be explained by chance.

"Second, one can test the various possible adaptive advantages of the feature, and its adaptedness is confirmed when all attempts to disprove these advantages are unsuccessful." Confirming or disproving why the brain enlarged is as inconceivable as confirming or disproving that hard thinking by Homo habilis to make tools jump started the jumbo brain. Another thought is the big smart brain evolved by a different mode and wasn't selected to promote survival and reproduction.

Some scientists think big brains come with big babies typical of humans. Their large size results from fast fetal development that continues long after other primates stop growing. A human baby is twice as big as the average three-pound gorilla baby that can grow into a five hundred pound silverback. Falk says the large cranium of human babies isn't caused by a higher degree of encephalization (brain growth), "but of a disproportionately large neonatal size." That is the big brain and big baby grow at equivalent rates during pregnancy.

What's more, the vascular system that develops during gestation is a component of the big brain that comes with the big baby. Still, it's not known if the big brain was simply a tagalong with the big baby, but it does act like a tested adaption. The adaptiveness is judged by people's proliferation and exploitation of natural resources, often to their detriment and other species. On the other hand, if the economic system that emerged in the Agricultural Revolution continues contributing to environmental ruin and drives people to extinction, the hefty brain would seem to be nonadaptive.

However, to label the brain nonadaptive, scientific consensus would need to specify the time a species must exist to be successful. Since what measures success is debatable, and nobody would be around to debate, it's a moot point. Gould says the largeness of the human brain is unrelated "to the demands of our body." Needed or not, brain growth was facilitated by prime releasers and prime movers. Prime releasers remove constraints allowing evolution to occur and prime movers cause evolutionary events. The cranial blood network of Homo habilis was the prime releaser that removed a constraint that permitted brain expansion.

The presumed prime mover of Habilis brain expansion was thinking in addition to other potential movers such as hunting. Other movers in the Homo species theorized to advance brain size to its present size were work, warfare and the vaunted capacity for language. Prime releasers and movers affecting the brain plausibly began when the ancestral apes that left the trees. Still, brains didn't expand initially, and scientific consensus is the disembarking apes adapted easily to the ground without big brains. Supporting the consensus is chimps are about as accomplished landlubbers as gorillas.

The chimpanzee "spends a large part of its time on the ground," Johanson says, "living on grubs, termites, berries, insects, buds and roots when the figs aren't ripe." Theoretically, apes living part-time on the ground could be stood up permanently by natural selection. The time needed to become completely bipedal would depend on the chance variability of favorable genes. The best genes for selection might not appear and be substituted by average or inferior variation. The alterations Nature makes on animals changing adaptive zones are graded excellent, good, fair or poor, and based on extinctions, failure.

Grading adaptions on a curve such as the human body and the mother eating beetles could raise or lower the grade. Adaptions "are often dictated by constraints and the availability of genetic variation," Mayr says. Natural selection prospers with ample genetic variation to alter all or part of an organism such as human hip muscles. The muscles support the pelvis for bipedal walking and were modified from the climbing muscles of apes. Johanson says if people decided to return to the trees, "these muscles could not return to the former way of functioning that relieved stress on the femur."

Returning to the trees to be an ape again would also require shrinking the brain and reducing its disproportionate areas.

Truly, leaving home was easy for the knuckle-walking apes, but going back would require evolution to reverse the adaptations that made them into bipeds. Although it's not macroevolution, Leakey says shifting "quadrupedal to bipedal locomotion demanded substantial changes in the body's anatomical structure." Restoring a biped to a quadruped and redeeming a small and rounder brain to boot would be an incredible reversal of evolution. As Thomas Wolfe said, "You can't go home again."

The main requirement for extensive changes in evolution to occur, Gould says, is that "organisms maintain a large store of genetic variability." Consequently, a round trip from ape to human would require gene selection to precisely retread all the roads paved by genetic variability. Moreover, people would have to retrace five million years of environmental shifts to be apes again. Backtracking through the large store of genetic variability is enough to prevent people from returning to yesteryear. It would much easier for the fanciful Incredible Hulk to turn back into Dr. David Banner.

Turning people back into quadrupeds is more fantastic had the apes left the trees on gimp legs caused by a tolerable mutation. Gimp legs are discredited by a theory proposing bipedalism was adapted for a new environment. Still, backtracking people into apes makes it easier to travel elsewhere or depart to extinction. A few genes that modify a few parts are often sufficient to convert a progenitor into a direct descendant or a species into a subspecies, but not vice-versa. A few genes easily turned chimps into bonobos and apes into bipeds. Genetic differences between humans and chimps are negligible.

Albeit changes from a progenitor into a direct descendant are noteworthy, changes during the evolution of animal phyla and classes are enormous. An immense revision on the class level of macroevolution began with the basic vertebrate genotype of the ancestral fish a half billion years ago. New genes were recruited and old genes recalled to fashion frogs, rattlesnakes, eagles and grizzly bears. New classes and species such as mammals and Asian elephants arise at irregular paces during micro and macroevolution because tempos of altering environments vary and optimized genotypes resist change.

The resistance is due, at least partly; to normalizing selection that Mayr says "eliminates all of those individuals of a population who deviate from the optimal phenotype." Still, in responding to natural selection, changes are slow or moderate or quick. Changes can be slight or drastic with the puzzling exceptions of living fossils like the coelacanth that has been static for seventy million years. Nonetheless, a living fossil in stasis could begin to modify by an environmental change or migration to a new adaptive zone or a mutation. A slowly evolving organism might modify faster for the same reasons.

'Living' fossil

Coelacanth

Other determinants of evolutionary rates are restricted gene flow that induces rapid speciation and uniform landmasses that slow down evolution in widespread populations. Regardless of the pace of evolution, most organisms meet extinction or become new species. A new species of long ago that met extinction or turned into another species was the aforesaid Stegodon, common ancestor of African and Asian elephants. Tracing the elephant lineage from the Paleocene Epoch sixty-five million years ago, Chadwick says, "The best paleontologists can do, is theorize a generalized marsh-dweller roughly the size of a pig.

"As the elephant evolved, the trunk, formed by a fusion of the upper lip, palate and nostrils gradually lengthened over time. Possibly, this organ made it easier for the animal to gather submerged vegetation while moving along the shores of a marsh and through shallow water. In that respect, it could be viewed as a unique alternative to developing a longer neck – – an alternative that enabled the animal to keep its head high enough to spot potential danger as it fed." Theory holds elephants were shorted long necks because of inadequate genetic variation or long necks weren't needed.

Instead of long necks, elephants adapted height and a flexible trunk to gather leaves from treetops. The elephants within a population with the best traits for the new environment were selected. The chosen ones became parents that passed their genes to their offspring for serviceable traits such as a better trunk and sturdy body. Their progeny grew to maturity and in turn produced offspring in the next generation with good traits that perpetrated the evolving lineage. Clearly, the long trunk and short neck are adaptations that have efficiently served the elephant lineage.

Giraffes, on the other hand, took another path in responding to environmental challenges and went with the lengthy neck. Perhaps the elephant's ancestor would have fared better with a long neck instead of a stumpy one, but didn't have a prime releaser to enable the adaption. Stumpy necks don't appear to be progressive or artistic like a red rose or a graceful gazelle. Roses and gazelles are usually pictured as "a process of inexorable improvement in form," Gould says. "Animals are delicately 'fine tuned' to their environment through constant selection of better-adapted shapes."

Generalizing, since fine-tuning sustains survival and reproduction, and elephants and giraffes survive and reproduce, they must be fine-tuned. Still, they are no more efficient in surviving than the tricky clams that mounted fake fishes on their rear ends. Evolution does not consider how animals look or act, but only how well they survive to parent offspring that reach reproductive maturity to shuffle along those 'selfish' genes. Many fine tuned elephants became work animals and bearers of maharajas after the Agricultural Revolution and others involuntarily die to tender their tusks for ivory.

Some giraffes and elephants wind up in circuses to amuse and perform for the not so finely tuned Homo sapiens. Untuned ancestral giraffes and Pliocene apes that became bipeds were faced with gravitational constraints. The elevator hypothesis further proposes that animals facing gravitational constraints in a new adaptive zone tune their bodies to balance the restrictive force. Gravitational stress on humans isn't problematic since their bodies were tuned through several precursor hominids. Yet, a reminder of gravity's constraint on the pioneering bipeds is sometimes felt when a person gets dizzy by standing up too quickly.

The giraffe lifted the gravitational restraint on its vascular system and selecting a twenty-five pound heart to pump blood up the long neck. When a giraffe lowers its head to drink, vessels and valves that were altered by its ancestors reduce the blood flow to the head. Similarly, gravitational constraints on "bipedalism necessitated a rearrangement in cranial blood vessels," Falk says. "The vascular systems of the gracile australopithecines that gave rise to Homo became modified in response to both gravitational and temperature pressures associated with refinement of bipedalism on the savanna."

The ancestral apes that gave rise to the first bipedal apes began tailoring their vascular system to counter the pull of gravity when they landed on the ground or soon thereafter. Perhaps they were experimenting with terrestrial living by commuting from the trees to the ground. This being so, the vascular adaptation could have occurred during the back and forth transits. Since gravity challenged the apes, natural selection dealt with it before dealing with thermal pressures that challenged the hominids in open equatorial habitats. First things first, for after all consensus maintains natural selection does not anticipate future challenges.

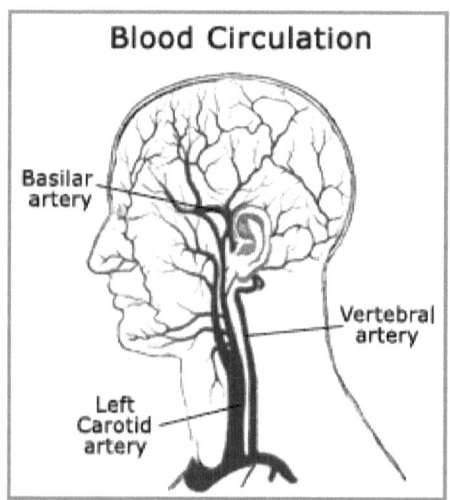

Modified Cranial Vessels

CHAPTER 12

Theory holds natural selection started human evolution in the Pliocene Epoch when descending apes began removing gravitational impediments. Subsequent alteration of crucial vessels into a novel blood network produced a prime releaser, and though it was needed, it wasn't the prime cause of brain expansion. Some gracile Australopithecines emerged from the stand up apes and moved from their shady habitats to hot savannas. The emigrant bipeds, in response to thermal stress, cooled their brains by modifying pertinent cranial vessels. The second customizing of the vascular system produced the prime releaser for one or more prime movers to enlarge the brain.

A core in the nutshell of human evolution is the series of inconsistent jumps from small to large brains, or A---B-C. The size increase is not a trend controlled by universal laws such as event A, without exception, is always followed by event B. One universal law, for example, is Newton's third law of motion that states for every action there is an equal and opposite reaction. Newton's physical law is demonstrated when a person falls back while roller-skating. When the person falls, the feet scoot forward and the upper body falls backward.

Physical laws and "mathematical generalizations can often be applied to biological phenomena like the Hardy-Weinberg equilibrium relating to the distribution of alleles in populations," Mayr says. "By contrast, all so-called evolutionary laws are contingent generalizations, and thus not equivalent to the laws of physics." An evolutionary 'law' states natural selection reacts to necessity, and although unproven, Darwinians accept it as fact. Modern Darwinian Theory holds the results of selection depend mainly on chance genetic variation. An imprecise generalization is Cope's law on the increase of body size such as the early reptiles gave rise to huge Mesozoic dinosaurs.

Dinosaurs "had a remote ancestor that was quadrupedal, walking on all fours" Milner says. "But the immediate ancestors of all dinosaurs were upright bipeds: the small, swift running thecodont. Some dinosaurs like tyrannosaurus continued to evolve as upright bipeds, while another lineage-including giant herbivores like brontosaurus-returned to being quadrupeds." Some descendent animals resemble their ancestors and some resemble intermediates while others don't resemble either one. Flying birds adapted wings that resemble the wings of their progenitor, an extinct flying reptile. Still, it doesn't mean evolution reversed exact paths to tap the wing genes from the reptilian genotype.

Evolutionary reversals are easily implied by 'degenerate' parasites that simplified their bodies for dependent lifestyles. However, they didn't backtrack, but selected new genes while mustering old genes for useful parts as natural selection gradually eliminated expendable components. There is no credibility that a chain of DNA in a genotype that has undergone even one alteration will ever have the link restored to its original location. A rule known as the Law of the Irreversibility of Evolution is Dollo's law. The law states that organisms don't turn back into a progenitor or degenerate by repeating events leading to speciation.

Dollo's law signifies Neanderthals weren't able to turn back the clock to their precursor species to avoid extinction. Dollo's law bodes Homo sapiens are stuck as people unless they evolve into a new species or meet extinction. Portending Homo sapiens are stuck in evolutionary limbo unless they change or vanish is a rule of thumb and not a universal law. Universal laws are better suited for the physical sciences since evolution is random. A selection event doesn't lead inexorably into a predictable episode. Rules of thumb apply to evolution like a population's unpredictable response to environmental change.

During an environmental change one individual in a responding population might react differently than another individual. "Different genotypes within a single population," Mayr says, "may respond differently to the same change of the environment." Regardless of how they react, organisms best suited for the new environ are favored to survive and reproduce. Some others that aren't as well equipped might barely survive, but fail to reproduce. "Owing to unequal survival and reproductive success of its individuals," Mayr says, "there is a continuing genetic turnover in each population as a result of chance and natural selection."

Surely, some members of a population are more successful than others, but it is a rule of thumb and not mathematically predictable. Falsely applying the rule of thumb to Newton's third law of motion means the skater falls backward, but the skate doesn't scoot forward. The skate, by a phony thumb rule, scoots off in any direction. An inflexible physical law and rule of thumb can be confused in situations that may seem predictable and sometimes unpredictable. A species, for example, is predictable to strive to survive and reproduce, but it can't be predicted that it will succeed.

Competing to survive is deterministic and difficult, but the struggle can be relieved by chance. An epidemic, for instance, may cut down on predators and allow their herbivorous prey to multiply. Herbivores typically multiply during seasons of optimum vegetation, which furnishes their predators more prey. Diseases, arms races and fluctuating resources govern cyclic successes or survival failures. Plentiful resources have helped most modern Homo sapiens survive since malnutrition has been their severest problem. Still, their unchecked proliferation and the competition for diminishing resources could sabotage many people's lavish lifestyle. Still, no matter how lavishly they live, their fate is guaranteed.

Scores of people are obese and out of shape from lack of exercise and overindulgent living. Overpopulating and gluttony are not unprecedented in mammals since some, if unchecked, will overpopulate and eat themselves to death. Some animals propagate to a point they have nowhere to expand. Animal populations expand by following the evolutionary decree of adaptive radiation that emerging species exploit and populate new niches, as did the Pliocene bipeds. Soon after the bipedal apes appeared they branched into untapped areas where they thrived and multiplied, eventually giving rise to the bipedal Homo hominids.

This first stage of hominid evolution began "when an apelike species with a bipedal, or upright, mode of locomotion evolved," Leakey says. "The second stage was the proliferation of bipedal species, a process that biologists call adaptive radiation. Between 7 million and 2 million years ago, many different species of bipedal ape evolved, each adapted to slightly different ecological circumstances." Since there are several possible responses to a new environ, bipedal apes moving into diverse habitats could have selected coincidental adaptions. Adaptions selected for an immediate objective are called ad hoc adaptions.

Conceivably, cranial vessels were ad hoc adaptions for gravity and cooling that were modified somewhat differently in separate domains. The slightly dissimilar vessels were analogous to one or two horns or humps on Rhinos and camels, both structures that were selected by responses to environmental challenges. Certainly, two humps on a camel weren't chosen to be a saddle for Bedouins or rhino horns an aphrodisiac for sexually dysfunctional Orientals. Along the same line, the purpose of a vascular alteration to purge gravitational and thermal constraints was not to produce a prime releaser for a prime mover to expand the brain.

Indeed, considering the radiator and elevator hypotheses are unproven, perhaps there was no selective determinant in changes of the cranial network and they were simply adjustments for vertical posture. Whatever revisions allowed or caused the brain to enlarge, they were characteristic solely of the Homo lineage. Many scientists think selective pressure forced Homo brains to expand consistently or in periodic spurts during their two and a half million-year evolution. A theory of brain enlargement holds the 1,350 cc Homo sapiens brain that has been static since the species emerged, took a flying leap from the 1,000 cc Homo erectus brain.

The leap to a big brain ties in the Out of Africa Theory that Leakey says proposes humans "originated as a single evolutionary event-a speciation-in a geographically discrete population." Conversely, the Multiregional Theory poses that Homo sapiens evolved from all worldwide populations of Homo erectus whose 1,000cc brains increased by 350 cc. Their brains were forced to swell by selective pressures activated by prime movers that emphasize intelligence and other superior qualities. Prime movers pressed and forced the survival of clever hominids while natural selection eliminated the less resourceful.

Nonetheless, pressure or force is "strictly metaphorical," Mayr says, "and that there is no such force or pressure connected with selection, as there is in discussions in the physical sciences." All members of a population undergo 'pressure' and 'force' that pressures and forces those that don't cut the mustard out of existence. The pressure and force exerted on populations of early Homo hominids presumably produced smarter individuals. The hominids passed their 'smarter' genes on to the next generation while the weaker genes were bred out or perhaps stored with other alleles with no obvious purpose.

Homo erectus grew bigger brains because they were pressed and forced by prime movers like needing to be better tool producers, hunters, warriors, workers and communicators. Perhaps extinction or pseudo-extinction blessed Homo erectus because that kind of intense competition might have made their lives miserable.

Initial brain expansion was conceivably caused by an event without pressure or force exerted by a prime mover. Leakey says brain growth was triggered by a dietary change expressed salubriously by you are what you eat. The Expensive Tissue Hypothesis poses that brain development and maintenance requires more calories than other tissues and organs.

The human "brain constitutes a mere 2 percent of body weight," Leakey says," "yet consumes 20 percent of the energy budget." Further, there was metabolic restriction on the energy needed for excessive brain growth that was balanced by a physical reduction in other parts. The human gastrointestinal tract is relatively smaller than a chimp's and as brains expanded the tract was reduced by natural selection putting on a balancing act. However, it can't be said the smaller tract wasn't caused by other modes like regulatory genes affecting embryonic growth and producing an offbeat phenotype.

The human body is quizzically topped by a big brain that needs lots of energy to operate. The brain's metabolic need is 400 calories for an average size person with a daily intake of 2,000 calories. Meat is rich in protein and fat and is highly caloric compared to plant food. Leakey says by eating lots of meat early Homo built "a brain beyond australopithecine size." Discoveries dated at brain gain time of chipped stones with sharp edges and fossil teeth that appear to be adapting to an omnivorous diet point to a meaty menu for early Homo hominids.

Falk says early hominids "sometimes left their own stamp, in the form of cut marks, on the bones of animals they butchered with stone tools." Since brains expanded after bipedal apes diverged, it's theorized that some apes became meat eaters by adapting mostly to the ground. There they became predators and scavengers depending more on meat than plant food such as figs. The meat eaters evolved into Homo hominids while the Australopiths remained semi-aboral vegetarians. Moreover, gravitational and thermal releasers of cranial vessels in one hominid group were possibly incidental constructs or by-products of another adaption.

This being so, the releaser was perhaps the only one conducive to brain growth regardless of the quantity of meat consumed. In this case, the gargantuan brain was the outcome of a small and incidental event in human evolution. The brain would hardly be a footnote except for the trouble it causes its possessors and other species. The human brain has remained static since humans evolved, implying people are smart enough and the brain doesn't need to grow any larger. Mayr says "in an enlarged, more complex society, a bigger brain is no longer rewarded by a reproductive advantage."

Moreover, no evidence exists of consistent brain expansion in the Homo lineage. Homo erectus fossils, for instance, show their brains grew in spurts, which refute steady brain expansion in that species. Perhaps inconsistent spurts denote Erectus needed periodic increases in intelligence to help them survive and procreate, thus passing on their genes. Organisms hustling to survive and disperse genes through phenotypes that inevitably die reflect the theory that genes use the bodies that house them as throwaway survival machines. Gould says self propelled genes are like a chicken that is "the egg's way of making another egg."

On the flip side, as humans propagate profusely and disperse abundant genes, it suggests they don't need to further enlarge their brains to survive. Certainly, a practical and introspective brain capable of solving hitherto insolvable problems would be nice. What's more, the skyrocketing global population, coupled with the bellicosity inherent in competitive societies bodes people should consider; limiting reproduction, declare a moratorium on the activity, or better yet, give it up completely. Socrates said pleasure and pain tread on each other's heels. Paraphrasing the Greek, on the pleasurable heels of intimacy, steps the pain of overpopulation.

Stepping hypothetically on the heels of a dumb Homo erectus that was devoured by ravaging hyenas was a crafty Erectus. The foxy Erectus outwitted the beasts and lived to tell, or garble the scary story to the children. Natural selection observed the hominids and rejected the dumb one condemning it to die. The crafty one was approved to survive and pass on its genes, perhaps genes for bigger brains. The two hominids were competitors in the struggle to survive, and, like other members of the population, had poor, fair, good and excellent IQs that were products of prime movers.

Deductively, natural selection picked the brain of the crafty hominid regardless of the size of the brain. Human intelligence isn't affected by brain volume and it's inferred that brain volume didn't determine the IQs of the extinct Homo hominids. Convolutions and brain size relative to the spinal cord size are supposedly just two of several factors determining IQs. Gould says variable brain volume isn't "extended to differences between species and certainly not to ranges of sizes separating ants and humans." All being equal, small or large brains interacting with prime movers are equally adept at helping people be successful.

The human brain, the gills of a fish, the wings of a bird and the croak of a frog are adaptations. Darwinians agree natural selection is the source of adaptions and evolution is gradual. Darwinians discount intelligent design, but recognize exceptions to natural selection such as genetic drift and sexual selection. What's more, evolutionists consider a novel event could have initiated rapid brain growth in the Homo line. Gould says natural selection makes no statement about evolutionary rates and many molecular researchers believe much change "is not only uninfluenced by selection, but truly random in direction."

The exceptions to selection and gradualism such as genetic drift and the speedily ballooned brain inspired the Omission Hypothesis. The hypothesis poses that the expansion of the human brain was neither adaptive nor a product of selection, but a result of an undetermined process. To confirm a feature is adaptive such as the expansion of the human brain is to test its advantages. When the advantages cannot be disproved, the feature is said to be adaptive. Mayr says it's almost impossible to substantiate that "any property of an organism is not of selective significance."

Proving the omission hypotheses that the disproportionately enlarged parts of the human brain aren't adaptive would be difficult. Regardless of no proof, the brain is significant since is fosters the maintenance of Homo sapiens. The brain is a larger, lopsided variant of the chimpanzee brain. Falk says both brains "appear superficially to have the same parts." The brains of the two primates are at the top of the IQ heap in the mammalian world. True, elephants and whales are smart, but their larger brains simply manage the activities of their huge bodies and do not award them the highest IQs.

Brains function to meet demands of an animal's subsistence in its environment and are not a teleological (designed with purpose) phenomenon. Gould says animal brains have no "preordained or intrinsic tendency for increase during the course of evolution." As most animals grow, their brains enlarge two-thirds as fast as their body weight. The general rule is the brain size of small or large animals is relative to their body weight. Raising a squirrel to the size of an elephant, for example, gives the squirrel the bigger brain, but the big squirrel's IQ would be the same as the little squirrel.

Conceptualizing the size of the human brain, Gould says "we must compare it with the expected brain size for an average mammal of our body weight." Gould estimated the 'average' brain weight of a mammal from the adult brain and body weight of many mammalian species. Chimp brains average 395 grams (cc) and Gould's equation shows an average mammal of the same body weight of a chimp 'should' have a 152-gram brain. Dividing 395/152 = 2.6 gives the encephalization quotient, meaning chimp brains are 2.6 times as heavy as the average mammal brain, and presumably 2.6 times smarter.

Quotients more than 1 stand for brains larger than average, and less than 1, smaller than average. The human encephalization quotient is 1,350/152 = 8.8 or a human brain is 8.8 times heavier than the brain of the average mammal, and evidently that much smarter. The brains of modern herbivores and carnivores are shown by their fossils to have increased through time, but it doesn't imply a teleological course from tiny to small to larger. Consensus is their brain size and intelligence increased from the competitive pressure in the constant arms races, wherein carnivores always stayed a little ahead.

The encephalization quotient of modern carnivores is 1.10, and compared to the calculated Australopith quotient, it suggests big cats were easily outsmarted by Pliocene hominids, even the most oafish. The 1.10 carnivore brain next to the 2.6 chimp brain shows why chimps are able to outsmart predators. Since humans emerged, their brain size with uneven hemispheres and assumed high IQs have remained the same. The high IQs were above and beyond the call of survival and procreative duty. Fossils imply the cerebral surge that led to the bright 1,350 cc brain began with Homo habilis, the first Homo hominid.

The surging brain that brought crowded teeth, troublesome sinuses, impacted molars, descending larynxes and other anomalies needed a bigger skull. To protect and house the bulging brain Gould says humans selected a bulbous, balloon-shaped head "unlike those of any other large mammal." Solving the riddle why people wound up with a globular dome topping a curious torso is like guessing on a multiple choice test. One untested answer is the bulbous cranium evolved independently and lugged the brain along like a car pulling a trailer. Then too, mosaic evolution might have enlarged the brain unevenly like the horse's skull.

Animals in evolutionary transition do not change uniformly," Milner says, "but by gradual degrees. Sometimes one part of the system may remain stable over long periods, while another evolves rapidly." Mosaic evolution changes animal parts drastically, moderately, little or none and parts that aren't targeted may trail along for the ride or lose their identity. Animals have parts whose functions were intensified or changed during speciation and barely touched parts that rode along like passengers in the back seat. Transitional animals descending from their precursor adjust specific components of the anatomy for new or similar roles.

"Any change of function simulates a saltation," Mayr says, "yet it is actually a gradual population change. It affects at first only one individual within a population and becomes evolutionarily significant only if it is favored by natural selection and spreads gradually to the other individuals of the population and then to the other populations of the species."

A moderate intensification of function, for example, that spread throughout precursor populations of giraffes was an extension of the neck. Giraffes have seven neck vertebras that are constituents of the basic vertebrate phenotype.

Throughout untold generations, transitional genotypes added new and co-opted old genes to lengthen the neck until it measured as much as six feet. The neck hoisted the head two stories above the ground letting the animal feed on lofty leaves. The giraffe's head wasn't the primary object of selection and necessity didn't require it to enlarge proportionately to the vertebra, otherwise it would be enormous. Giraffes with the longest necks survived better to pass on their genes. "Those individuals that are best adapted to the abiotic and biotic environment," Mayr says, "have the greatest chance to be among the survivors."

Giraffe

Stuck out its neck

CHAPTER 13

Conceivably, one of the hominid lines diverging from the ancestral ape entered the harshest environment and natural selection began enlarging their crania for some undetermined reason. This being so, it follows as mosaic evolution distended the skull in the Homo lineage the cerebral hemispheres tagged along. Proving the extending crania dragged the brains with their added bulk with them and the process wasn't a process of selection, nor was it adaptive, would confirm the Omission Hypothesis. The proof would enlighten evolutionists and be unacceptable to intelligent designers.

Proving that two-thirds of the human brain wasn't an adaption suggests labeling the extra gray matter a benign tumor. A tumor is medically defined as an abnormal growth of tissue caused by an uncontrolled, progressive multiplication of cells and serving no physiological function. However, in evolution, structures are neither abnormal nor normal because events are random ad hoc adaptations. The extra gray matter affords more intelligence like a big muscle is stronger than a small one. The human brain has utilized accumulated knowledge passed through generations to build a world of massive productivity and technological inventions.

The particles comprising the brain that causes much unrest were present in the Precambrian seas rife with quasi life forms. The minuscule forms evolved into unicellular organisms that gave rise to multicellular animals that diverged into numerous bushes. Fossil evidence shows that two twigs on one bush budded Homo habilis and Homo erectus with waxing brains. On an outer twig like the last rose of summer Homo sapiens was seeded with static brains. Promoters of intelligent design behold the stationary brain to be the final goal of a linear progression called the Scala Natura or Great Chain of Being.

The belief in linear progress "unfolded long before evolution became a familiar thought in our culture," Milner says, and "dominated the Western view of nature. Everything in the universe was arranged in a hierarchy, from low to high, from noble to base." On top of the hierarchy people are perched in transcendent glory while all other creatures are hooked on descending links of the Scala Natura. Spiritual beings hover above the chain of humanity with an immortal roosting at the zenith of the universe. Vertical chains of moderate to despotic hierarchies have existed in human societies throughout recorded history.

Roosting atop the Universe

A theory poses the chains of ranked systems began in the economies that arose during the Agricultural Revolution. Kings, lords and serfs exemplify degrees of human hierarchies from top to bottom. Military hierarchies range widely from the supreme commander to the lowly foot soldier. Subtle hierarchies are seen between husbands and wives when it's uncertain who manages the marriage. Human hierarchies are analogous to the hierarchal society of chimpanzees like the Gombe community studied by Jane Goodall. Some researchers think the ape hierarchy at Gombe developed during the corruption of their adaptive social structure.

This "hierarchy is not the normal form of organization of chimpanzees," Power says, and "the alpha Gombe animal is a despot, using his power oppressively, rather than serving the group as a protective leader." Chimpanzee males have the adaptive role of general protector, but a despotic male Goodall mentions, except for one good deed, never assumed a protective role. Clearly, a bipedal youngster with the character of this selfish ape would hardly grow up to be a scoutmaster. Low ranked apes are in awe of dominant apes like intelligent designers are in awe of the mystical maker of humans.

The designers imagine Homo sapiens are peerless and human perfection denies error its existence. However, the imperfection of human 'wisdom' is revealed in countless wars throughout history that continually kill people. Some scientists believe human aggression and warfare was ingrained in the genotype of the ancestral ape. They base their belief on the observation that warfare among nonhuman primates is not without precedent. Indeed, in the Primal War between the northern and southern Gombe chimpanzees, their smaller roundish brains didn't prevent the northerners from killing their former friends from the Southland.

Goodall says the northern apes began a southern campaign "that culminated with the annexation" of the southern range and "the complete annihilation" of those ill-fated southerners. Clearly, and philosophically, error denied perfection and the southern chimpanzees their existence. Perfection culminating through evolution into a flawless adaption, like beauty, is a matter of perception. Beautiful flowers bloom merely to wither away like species rise and thrive only to fade into extinction. A vision of perfection is to improve imperfect organisms until mortals rise to the surface of the Chain of Being like cream rises to the top.

Denying perfected humans, Gould says, "evolution has no direction; it does not lead inevitably to higher things. Organisms become better adapted to their local environments, and that is all. The 'degeneracy' of a parasite is as perfect as the gait of a gazelle." Additional degenerates "among the higher organisms," Mayr says, "are cave animals, subterranean animals, and other specialists that show many retrogressive and simplifying trends." Another perfection denied was bone weakness that limits growth of huge animals like Brontosaurus that was partially supported by water. The one hundred ton blue whale is completely supported by water lest it collapse.

Disappointing for science fiction fans, the leg bones of a real King Kong could not support his weight. Natural selection would need to make bones out of stronger material to build King Kong. Selection can build animals only so big and it can't frame a mammal smaller than a pygmy shrew since its body would lose too much heat. Terrestrial mammals needing phenotypes smaller than a shrew would require severe thermal adaptations. Nature does what she can with the chance offerings of genetic variation, and without purpose constructs endearing animals like rabbits and puppy dogs.

People who believe Nature perfected humanity through the Scala Natura, associate human perfection with love, compassion and generosity. Animals with certain traits that appeal to feelings of sentiment stir those emotions. Some pets often return more love than children and are much less trouble. Preferring pets to children questions the adaptiveness of directly competitive societies that emerged after the Agricultural Revolution. A typical feature of Western societies is a generation gap between parents and children, especially adolescents. Frequently cited as the gap's cause is the economic system of direct competition that polarizes the family.

Some current primitive societies apply harsh punishments like they were applied in ancient times. Old Hebrew laws demanded a stubborn and rebellious son be punished by "all the men of his city stone him with stones, that he die." (Deuteronomy 21:21)   Milner says, stressful social situations and hormones frequently add to the difficulty of adolescence in the affluent Western World. Babies are temporarily spared the difficulties since they don't have the hormones or the social stress, and they are cuddly.

"There seems to be a strong evolutionary basis for why we find the face of a baby cute, and by extension, the similar faces of baby animals, dolls and cartoon characters." Mickey Mouse and Teddy Bear were physically modified after they were introduced into "recognition patterns deeply imbedded in the human brain, a form of parental behavior that has helped our species survive. A certain kind of face seems to release protective, parental responses; we want to cuddle and care for helpless infants." Male chimps respond likewise Goodall says and often can't "resist reaching out to draw an infant into a close embrace, to pat him, or to initiate gentle play."

Surely, Nature 'wants' a baby to reach reproductive maturity that calls for all members of a society to cooperate. It takes a village to raise a child is applicable to the cooperative practice of hunter-gather societies. Plausibly, the cooperative instinct of raising children, so pronounced in humans and chimpanzees, is Baldwinized in their genotypes. The length of parental care varies among mammals and some offspring must fend for themselves soon after weaning. Cartoon characters like Mickey Mouse, pet dogs and cats, pandas and baby chimps in diapers evoke people's parental instincts.

The endearing panda is a bear that natural selection 'rigged' with 'contraptions' to help it survive. The giant panda feeds on bamboo, but its digestive tract isn't completely adapted for plant roughage. Therefore, the panda eats voluminously to sustain itself. Added to its quizzical phenotype, the paws weren't selected to pluck stalks, possibly because of insufficient genetic variation. Instead, selection rigged a 'thumb' from a small wrist bone to hold the bamboo shoots. Pandas are referenced in a theory that poses in cases of co-adaption or mutual dependence; change in one can promote change in the others.

Panda

Rigged to survive

The change may be harmful like the loss of an essential plant food of herbivores that causes their demise and a domino effect among other species. The panda is an animal that evolved with an inefficient digestive tract and a 'rigged thumb'. The thumb was selected because resources of its omnivorous precursor decreased quickly, leaving only fibrous bamboo. Animals are rigged to survive by favoring the best individual differences along with their genes. Selection doesn't wait for better genes, but if some appear during rigging, they too are employed. Rigging indicates natural selection is deterministic, but aimless and without purpose.

Characterizing animals as poorly adapted because they are rigged with contraptions is a visceral, but inaccurate evaluation. Outlandish animals satisfy Nature's determinant to survive as successfully as Peter Rabbit or Winnie-the-Pooh. Reality and what 'should be' such as a beautiful flying ostrich are not the same. Animal phyla contain related species whose phenotypes deviate from people's perception of the norm from delightful to grotesque. The panda is delightful and so is a nautilus with its beautiful shell. The nautilus is a close relative of the bug-eyed octopus with a soft body and eight tentacles.

Intuitively, an octopus is a frightful creature that often plays a villain in horror movies. Nautiluses and octopuses are cephalopods with dissimilar external bodies, but similar internal anatomies, which is true of people and chimpanzees. The underlying structures of humans are "remarkably similar to that of the chimpanzee," Mayr says, "one of the reasons why Linnaeus did not hesitate to place the chimpanzee in the genus Homo." Obviously, the conclusion of the Swedish naturalist to clump people with chimps was ignored by the sapient species wishing to eternally remain at the crest of the Chain of Being.

Octopus

Scary

Nautiluses and octopuses, like people and chimps, don't look like close relatives judging by external anatomies. The two cephalopods are at opposite ends of human perception of beauty and range from delightful to abhorrent. Their common ancestor was altered superficially during environmental variations while maintaining the internal cephalopod anatomy. The external anatomies of creatures such as spiders, flies, owls and pussycats reflect those of their precursor. The presumed reason is the habitats of predecessor and descendants were similar. These animals illustrate selection acting over countless generations favoring individuals best suited for the environment. Humans appear to be exceptions.

Proving the bipedal primate is an exception would be as improbable as proving extra brain matter isn't a product of selection. Most people couldn't be convinced they are anomalous apes because of cultural indoctrination. One anomaly that questions their position in the Great Chain of Being is the female mammary glands. Women's breasts vary in volume from a tangerine to a honeydew melon whereas the mammary glands of nonhuman primates barely grow after puberty. Size doesn't affect milk production of lactating women and science has no answer for volume variation. Presumably, breast and brain enlargement began with the Homo hominids.

One hypothesis for voluptuous breasts is they excite males to copulate and are favored by sexual selection. "Features are sexually selected when they are preferred in one sex by the other and are passed down to future generations," Falk says. "A classic example is the beautiful tail feathers of male peacocks. Apparently, males have beautiful tails because females prefer to mate with (and, therefore, bear the offspring of) males who are so endowed. Such 'flasher' theories have also been applied to the large breasts of human women, as opposed to their relatively unendowed nonhuman primate cousins."

Hypotheses centering on large breasts as sexual stimulants harmonize with eroticism prevalent in many circles of the Western World. The security of chemical contraceptives introduced in the mid 20th century influenced the rise of sexual liberation from the rigidity of Puritanism. Equating sexual passion with sin, Puritanism prescribes that pleasure not be reaped from the reproductive act. Victorian propriety had kept the female body under layers of clothes when an exposed ankle was seen as moral degradation. Exploiting sex for profit increased markedly with the decline of moral codes in the affluence of industrial nations.

Magazines, books, movies, television, and the Internet profit from sex, which includes child pornography. Commercialized sex glamorizes women by accentuating curvaceous legs and full breasts that are often surgically augmented. Adulterated breasts are skillfully sculptured and as hard to distinguish as butter from margarine. Ironically, surveys by men's magazines report large breasts or breasts in general don't excite all men. Some men prefer rounded rumps or fetishes like the shape of an ear or foot. Stimulation by breasts covered under clothing is plausibly goaded by innate curiosity, a characteristic of humans and chimpanzees expressed in the forbidden fruit syndrome.

Forbidden fruits

Loss of Eden

Surgically enlarged breasts

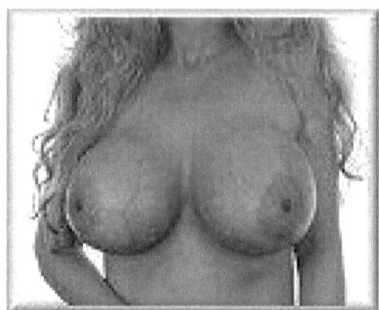

Exciting? Repulsive?

Provided sexual selection favored larger breasts over generations, women would be as top heavy as broad breasted turkeys that artificial selection misshapes. Possibly, women's breasts are benign tumors, like the hypothesized extra bulk of the human brain. One study of primitive tribes revealed the men are indifferent to breasts. The women don't cover their breasts and mothers nursing babies is usual. Symbols of feminine beauty in another tribe are flattened lips that look like pancakes, and to Western perception appear painful and bizarre. Moreover, intercourse among primitives is without the embellished foreplay touted in sophisticated societies of the Western World.

Average size breasts

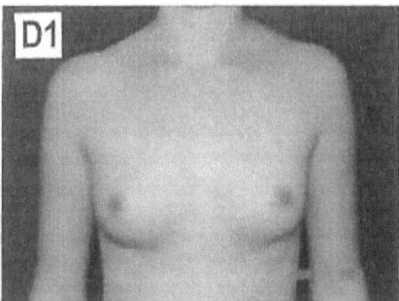

Benign tumors?

Sexual practices of remnant hunter-gatherer tribes infer the copulatory position of human ancestors was simplistic like mating chimpanzees. Sagan says primitive couples often pair up for sex "recumbent on their sides, the male embracing the female from behind." Conversely, in seeking pleasure within sophisticated societies experimental sex and extended foreplay are usually practiced prior to penile penetration. Hypothetically, when basic needs are satisfied, idle people indulge in unconventional sexual acts to fill the void with thrills. Gould says some erotic acts men and women practice on each other simultaneously or individually have higher incidences "among the upper classes."

One upper class group was composed of wealthy Roman families called Optimates that became egoistic and addicted to lavish living. Optimates were arrogant patricians or ancient 'swingers' who forsook traditional principles for debauchery and carnality. Quirky practices are not restricted to leisurely people lounging around the upper echelons of past and present flush societies. Domiciled chimps often experiment with unorthodox sex although not as sophisticated as whips, chains and black boots. Goodall says they experiment because they "are less concerned with the pressures of day-to-day living than in the natural state and have more time for nonadaptive experimentation."

Evidently, extended leisurely living may lead the two primates into temptation where they tinker with sex, or nurture and not nature. Credibly, cultural nurture coupled with societal orientation is the candidate responsible for augmented breasts and flattened lips. Thus, one person's simulative treasure is another's trash, which refutes sexual selection as the sculptor of extra large breasts. Variance in breast size is a puzzling anomaly possibly initiated by brain expansion retarding the endocrine system causing a hormonal imbalance during gestation. The imbalance wasn't uniform among female babies and individual differences produced variable breast sizes.

Flattened Lips

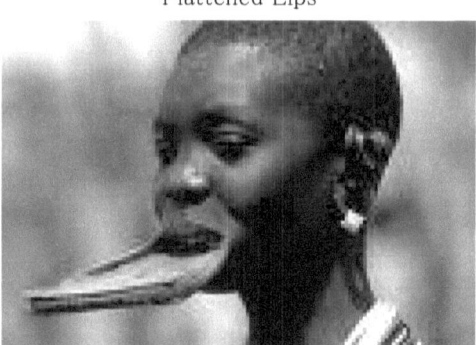

So lovely to look at

Variable breast sizes might be caused by regulatory genes, or the genes could induce unknown processes resulting in erratic breast contours. Whatever size breasts are to be, they begin growing in pubescent girls when the ovaries secrete the hormone estrogen. The breasts reach maximum volume by the action of the hormones corticosteroids, progesterone and prolactin. Although varying size is an aberrant human condition, it's not a disorder like gigantism or dwarfism caused by excessive or insufficient growth hormones. Discovering why breasts vary will remain a mystery unless researchers find causative genes or fluctuating size is a by-product of another condition.

There is a similarity between the titillating breasts of a woman and the pinkish buttocks of a female chimp in estrus. The crimson rump, so provocative to male chimps, is suggestive of the backward glance of an alluring woman rosy with lipstick and rouge. On the contrary, menopause is a striking difference between humans and chimps and other primates. Menopause or 'change of life' begins in middle aged women. The ovaries gradually stop producing ova resulting in diminished estrogen and terminating menstruation. Menopause and the lost vigor of the ovaries render women thereafter barren.

Estrus chimp. As provocative to male chimps as an alluring woman aglow with lipstick and rouge is to men.

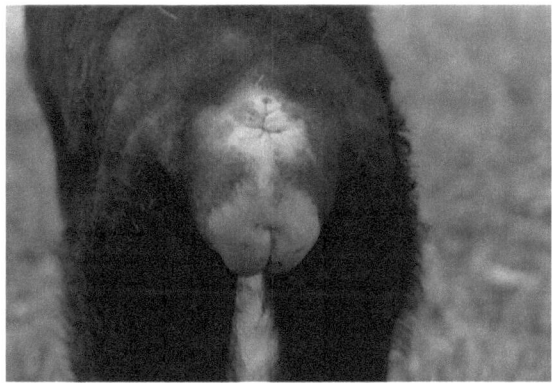

Artificial sexual stimulant of painted pinkish lips

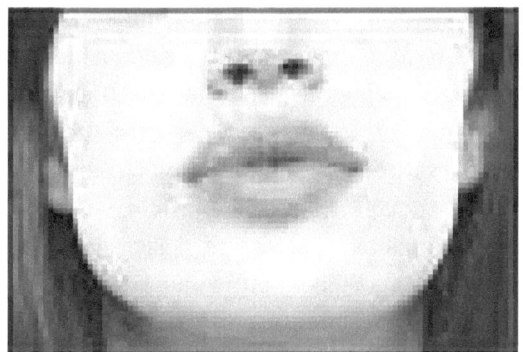

Like two titillating apples off the same luscious tree.

Ovaries operating at their peak set breast maturation in motion and gear women up for procreation. Although menopause and varied breast sizes are anomalous conditions that may negatively affect women's health, it cannot be proven they are not normal courses in human growth and development. Diamond says menopause was selected because of "the exceptional danger that childbirth poses to the mother" and "the danger that a mother's death poses to her offspring." Nevertheless, the decline of estrogen throughout menopause may cause physiological and psychological disturbances in women ranging from mild to severe.

Estrogen deficiency often leads to osteoporosis characterized by porous bones that frequently fracture in older women, especially the upper femur, erroneously called a broken hip. Johanson says most of the 'broken hip' injuries that afflict innumerable people "each year occurs in the femur, which lies next to the hip joint." Usually, the person falls when the femur breaks without more bones being broken. Menopause is one of many anomalies of women, but anomalies in men are just as pronounced, bringing to mind the old saying, "what's good for the goose is good for the gander."

Determining that the aberrations of men and women are adaptive would require evidence they are beneficial to survival and reproducing offspring that reach maturity. Conditions such as menopause and assorted breast sizes don't obstruct Nature's prime directive to survive and reproduce even though they can be problematic or inconvenient. Prophets of intelligent design don't consider the two conditions anomalies, but as improvements along the Scala Natura that move toward perfection at the vertex of the Chain of Being. The conditions serve the prophets to distinguish people from the other primates.

Opposing the prophets, are evolutionists who think menopause and varying breast sizes emphasize that natural selection is determined, but haphazard and unpredictable. Evolutionists believe anomalies are made with available genes or they are products of another process. Whatever is responsible for menopause and motley breast volumes, they were initiated by the ovaries. The ovaries are regulated by the master pituitary gland and other hormonal components located at the base of the brain. Credibly, when the brain started to distend in an early hominid the pressure on the rising skull floor mashed and permanently damaged the delicate pituitary structures.

Provided early glandular damage during evolution caused the problems of women's breasts and menopause they are nonadaptive or maladaptive. Nonadaptive and maladaptive also applies to animal groups such as the corrupted Gombe chimpanzees that were once peaceful. A feature that is nonadaptive or maladaptive may have performed optimally in a precursor, but marginally or poorly or not all in the descendant. Ostrich wings, for instance, seem nonadaptive, but they are also vestiges of ancestral wings adapted for flight. Another vestige is the human appendix that might be adaptive because its extra space can relieve abdominal gas.

Vestiges are inherited and harmless remnants that may remain nonfunctional and left alone by natural selection. However, they aren't heritable traits coded by a deadly or combination of mutated genes that natural selection quickly selects against. Pernicious characters that are severe in some cases, but not in others may or may not be eliminated. A pernicious ambulatory trait like clubfeet, for instance, might be lethal for a zebra and tolerable for an arboreal ape. "A character of an organism," Mayr says, "is an adaptation when among the variable populations of the ancestors it had been favored by nonelimination."

Mayr doesn't say how well the character worked, only that it wasn't eliminated. Noneliminated characters can be substandard traits that were allowed to remain. Unless ostrich wings are useful like balancing the bird, they are nonadaptive or neutral. Provided a trait is nonfunctional and

troublesome, it is maladaptive. A trait such as hummingbird
wings that functioned optimally in the precursor and
descendant is super–adaptive. When the function of a trait
was poor in both precursor and descendant it is sub–
adaptive. What definition befits the panda's thumb and
digestive system is arbitrary, easier viewed as a concept.

   Another concept is an adaption that is defined as a property
of an organism selection favored over alternate traits. Darwin
thought adaptions are never perfect, but come in grades from
poor to superior. "Individuals that do not have as good an
adaptation as others are eliminated," Mayr says, "but the
survivors do not contribute to the process of becoming
better adapted by any special activities, as proposed in
teleological theories of evolution." Ostriches, for example,
cannot run and flap their wings to fly followed by coding
genes for flight being ingrained in the genotype.

   The ostrich's ancestral genes for flight are plausibly stored
in the genotype with other unemployed genes. Nature picks
genes from genetic variation while others are co–opted old
genes and perhaps some are recalled from the storehouse.
'Timely' genes coded for the optimal reproductive strategies
of Emperor Penguins and Albatrosses. The big birds
reproduce late in life and lay one egg under harsh conditions,
but their longevity and few aerial or terrestrial predators
balance their discomfort. On the other hand, most birds such
as bluebirds live under better conditions and have several
offspring, but more predators and shorter lives.

CHAPTER 14

Natural selection is deterministic to promote the survival of organisms like penguins or bluebirds, but it cannot adapt them for comfort or prevent discomfort or nonadaption or maladaption. Selection can only use the available genetic variation for the survival of organisms. The concept of adaption, nonadaption and maladaption are like beauty, which is in the eye of the beholder. Beautiful adaptations are presumably acquired slowly in cadence with gradually altering environments. The rational is the more time natural selection has, the more beautiful the adaptations, as people perceive beauty.

Provided Nature works best gradually, it adapted the great apes optimally as they slowly evolved from their common ancestor. Orangutans adapted arboreal life and gorillas opted for terrestrial life while chimps utilized both niches. The three apes are closely related and their phenotypes are constructed on the basic primate genotype. The primate genotype was built on the mammalian genotype that was assembled on preceding genotypes back to Precambrian bacteria. "During embryonic development," Mayr says, "the basic features of the body plan are laid down before the special adaptations for their niches begin to develop."

During the development of the apes in the womb, mammalian genes are added to the basic vertebrate genotype. Once the mammalian characters are laid down, primate genes alter the mammalian characters and finally specific genes shape the designated ape. Constructing an ape from the mammalian genotype is analogous to building additions to houses like bathrooms and porches. Construction of the human primate from the genotype of a chimpanzee was done recently, which is not obvious by the phenotype. People aren't classified as apes, but in a devalued Great Chain of Being their DNA sequences categorize them with chimpanzees.

Measured by DNA analyses shows the first of the great apes to diverge from the common ancestor were orangutans. Moreover, as they diverged they selected adaptions to acclimate them to their new niches. Nature fitted the three quadrupeds with novel structures, physiological traits, behaviors and other useful features. Physically, orangutans, gorillas and chimpanzees are similar since they retained most of the characters of their precursor. The trio emerged separately by adding the genes of their novel characters to the genotype of their common ancestor, the unsung antecedent of human evolution.

Orangutans selected several novel traits including hook-shaped hands and feet to grasp branches and vines. They seldom venture to the ground, and when they do, they walk on all fours. Orangutans are solitary animals that use their modified appendages to move slowly through the trees foraging independently for food. Their diet is much like the chimpanzee diet, consisting mostly of plants and supplemented with bird eggs, termites and an occasional monkey. Orangutans are intelligent apes that use leaves in the wild to fashion umbrellas or fold into cups for water. Laboratory orangutans can select or fashion tools for future use.

Orangutans adapted a matriarchal social structure, but gauged by the gregariousness of their gorilla and chimp cousins, the Asian orangutans are loners. The orangutan male has exclusive control of a territory wherein several females occupy their own areas. When a female becomes fertile, she seeks out the lone male for a brief affair until she is pregnant and then they go separate ways. Occasionally, two mothers with youngsters pair up for a few days so the children can play. Playtime is about the extent of their sociability other than the brief liaison of an amorous couple.

Their solitary lifestyle is adapted for a habitat high in the trees that isn't the best place for big primates to make merry and hobnob with friends. Some scientists question their solitary behavior since rescued orangutan orphans seek the companionship of their fellows and are unhappy when returned to their natural habitat. Milner says their behavior might be a "necessity since the Malaysian forests cannot support great concentrations of these large apes in one area." Their preference for companions when they are orphaned and the gregarious character of the other great apes implies their common ancestor was a social ape.

Orangutans became well adapted to their environs by slowly building on their precursor's traits, but it doesn't mean the process was directional. Evolution is aimless and "does not lead inevitably to higher things," Gould says. "Organisms become better adapted to their local environments, and that is all." Regardless of how good orangutans are adapted, their bipedal cousins are pushing them toward extinction. They are captured for zoos, to sell as pets and hunted for meat, but their biggest threat is habitat loss from logging. Their eradication isn't deliberate, they simply aren't as important as the profit from harvesting lumber.

Foreclosed homes

Orphans in a sanctuary

Young orangutan with his old dog friend

Clearly, forcing animals to extinction causes many people to feel guilty because of their compassion, an adaptive property to sustain their original social organization. However, destruction is inherent in the human economic system that arose 10,000 years ago. In a nutshell, a bipedal ape's gain is a knuckle-walking ape's loss. The apes that gained plenty, at least in weight, are gorillas. Gorillas entertain people in zoos and movies especially in the trumped up reputation of aggressive apes. Fossey says gorillas are innately mild mannered and the "extraordinary gentleness of the adult male with his young dispels all the King Kong mythology."

Gorilla groups are composed of five to fifteen members; the huge silverback, several mature females, youngsters and one or two subdominant males that don't breed while in the group. The family life of gorillas is congenial and remindful of television's The Waltons. The patriarchal silverback is twice the female's size and his caretaker role evolved over thousands of generations by directional selection favoring specific individuals in a population. His position is demanding and often dangerous in the relentless surge of human encroachment on their habitats. "Uneasy lies the head that wears a crown." (Henry IV)

The big silverback emits a strong odor from glands that natural selection adapted to facilitate olfactory communication and terrestrial travel in their niche. Adaptions are "organized at different levels," Mayr says. "This makes a specialization for highly specific niches possible." Gorillas are efficiently adapted, but they are threatened because encroachment and poaching by their vertical cousins are reducing their habitats. Tensions and aggressive interactions are increasing among gorilla groups due to waning living space or Lebensraum. Crowding is disrupting the economics of their foraging system for which they have no adaptive response.

Indeed, much of Fossey's book is focused on their plight and hope for better conservation measures. Many scientists conclude gorilla societies were once as peaceful as cooing turtledoves before Homo sapiens entered the placid picture. The silverback strives to avoid conflict with other groups in their constantly dwindling habitats. Somewhat compensated for his effort, he has all the pleasures of exclusive breeding rights to his harem. The subdominant or nonreproductive males eventually leave to found their own group or loosely attach themselves to other bands. Subdominant does not denote subservience or inferiority that identifies people in nonadaptive social organizations.

Neither does subdominant or subservient describe females because adult relationships are not based on power, but mutual respect. The females are seldom related since when they reach adolescence they emigrate from their natal group to join a male. Some scientists consider female emigration adaptive to disperse genes that prevent concentrating their gene pools. Adult females emigrate to different groups by solicitation from other silverbacks or better breeding opportunities, and possibly for more security. An opinion is females leave their adopted groups vainly seeking security because human encroachment has corrupted their social structure rendering it nonadaptive.

Fossey says "less than a hundred men, armed with bows and arrows, spears or guns, have been allowed to plague the wildlife in the parklands that form the last stronghold for the mountain gorilla." Truly, or rather untruly, for an insecure female to reproduce she might have to be unfaithful to her silverback. 'Faithful' is not inherent in the polygamous nature of silverbacks that precludes exclusive relationships like those between monogamous gibbons or faithful men and women. The sage silverback holds the bonds that bind gorilla groups together, but when he ages, a younger male often supplants the respected primate.

Fossey says two maturing males on their own "frequently interacted with other groups as they sought to obtain females to establish their individual harems and, ultimately, their own group." When a younger male dethrones a patriarch, the elderly ape wanders lonely as a cloud bereft of fellowship and feminine pleasures. The image of an expelled silverback befits Milton's description of an angel ousted from heaven. "If thou beest he: But O how fallen! How changed from him, who in the happy Realms of Light-- dist outshine myriads though bright."

An old silverback compared to a fallen angel is an anthropocentric, but skewed view of gorilla society. A capable silverback is essential to gorilla groups, but eventually his energy declines and he peters out. However, Mother Nature provides a prepackaged adaption in his youthful replacement. Further, the old ape would hardly ache for feminine pleasures since his sexual equipment is the smallest of the great apes and below par by human perception. His testicles are 0.017 percent of his body weight and are tiny compared to a chimp's testicles that Goodall says "account for 0.269 percent of body weight."

The Waltons.

Placid group. The juvenile human endocrine system limits their sex hormones.

Thus, the immense testicles of a male chimp are more than four times bigger than a huge silverback's testicles, which seem upside down. However, nature doesn't cotton to errant impressions like testicular sizes or falsely assuming an eighteen-ton giant whale shark has bigger teeth than a runty two-ton great white shark. The irony of giant and runty testicles pertains to the theory that primate testicular sizes are adapted to operate in sync with their reproductive strategy. The silverback's tiny testicles are big enough to afford ample ejaculations for his infrequent breeding pleasure.

The big fellow has dry spells in his sex life since he services only the few females that are receptive for several days a month. When females give birth they remain sexually inactive for several years. An orangutan male's testicles are a little larger because he breeds slightly more than his gorilla cousin. Conversely, Goodall says chimp's testicles are colossal since they are adapted to "meet the high copulatory frequencies in a promiscuous society." Promiscuous is pejorative because it was sullied by puritanical hierarchies and connotes Victorian mores. Calling chimps promiscuous implies they are naughty and are doing something wrong.

Promiscuous people are indiscriminate in their choices of sexual partners, which is perilous in view of the HIV virus. Nonselective sexuality describes the reproductive strategy of chimps without a negative connotation. Another word is polygamous that is a concise and useful descriptive adjective, though it applies to simultaneous spouses. Polygamous chimps are sexually impassioned primates exemplified by an ardent old male who was seen copulating three times in less than five minutes. Gorillas and orangutans could not contemplate such a feat, which might kill any man foolish enough to give it a stab.

Primate testicles and penises were selected for their lifestyle to produce and deliver adequate sperm to assure impregnation. Mayr says the lifestyle of an organism is impressive by "the presence of very specific adaptations that make this lifestyle possible." Further, penises of the hairy great apes are in a measure correlated to the weight of their testicles. Goodall says the erect penis of a chimp is "8 centimeters long, compared to the 3-centimeter penis of a fully adult gorilla, the 4 centimeter penis of an orangutan, and the 13-centimeter mean length in the human male."

The relative body percentage weight to penis length of chimps and gorillas is 0.269/8cm and 0.017/3cm respectively. In a nutshell, chimps have long penises with large testicles and gorillas have short penises with small testicles producing much less testosterone. Reducing a man's body volume to the body volume of a chimp would make both penises similar in size. Curiously, the big, or rather small difference is, the man would have peanut size testicles. Put another way, a chimp the body volume of a man would have a comparable penis, but his testicles would be as big as grapefruits.

The length of the human penis implies it's adapted like a chimp's penis for frequent copulations in a nonselective sexual society. However, the chimp's penis is backed up by brimming testicles producing copious semen for multiple inseminations. "Natural selection," Goodall says, "has provided the male chimpanzee with an exceptionally efficient mechanism for production, storage, and serial ejaculations of viable sperm." Conversely, a man's testicles aren't in the same ballpark with a chimp's testicles. Ridley says men's sperm production is inferior and their runty testicles don't "operate on full power (that is, they might once have been bigger in our ancestors)."

The runty testicles might be an ad hoc or for the moment adaption for some ancestral dilemma that was later resolved. Certainly, Mayr says, an organism has to be well adapted to survive, but it must be able "to cope with its ancestral genome. Not every part of an organism is an ad hoc adaptation for its present lifestyle." Provided humans descended from large testicle chimps, a heritable genome plausibly should code for much larger testicles, which it doesn't. Perhaps the substandard testicles were made by intelligent design to curb population growth or they are simply sputtering maladaptions.

Conceivably, coding for potent testicles in a past Homo hominid was disrupted by a mutated regulatory gene affecting embryonic development. A mutated gene shrinking testicular size denotes it integrated into the genotype and forever stunted testicular growth in the Homo lineage. The shriveled testes were credibly one anomaly initiated in Homo habilis by the swelling brain squeezing the pituitary gland on the skull floor. Hence, the testicles of all Homo habilis posterity shrank in step with the ballooning brain of each new species. The last testicular squeeze in the Homo lineage left men holding the smallest bag.

The idea of mashed hormonal tissues assumes the precursors of Habilis had long penises and big testicles chosen for a polygamous society, but proof is unlikely since decaying flesh doesn't fossilize. Clearly, a man's long penis can't perform in a series of quickie copulations in the manner of his prodigious cousin. Natural selection either didn't have time to trim the long penis down in proportion to the small testicles or there is no adaptive value in pruning the organ. Obviously, if deep penetration to spill sperms near the Fallopian tubes abets reproduction, cropping the penis would be disadvantageous.

Pituitary gland

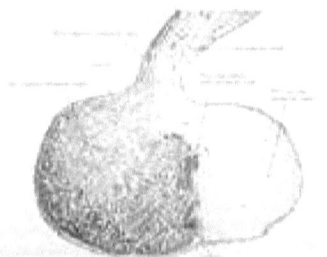

squeezed on the skull floor??

Increasing the body size of a male chimp to the body
size of a man shows a comparable size penis

Another hypothesis is lengthy penises were picked to foster
fertilization since the cloistered slot of a woman's vagina
doesn't optimize penile penetration. The reclusive vagina is
neither fore nor aft inasmuch as it doesn't rotate rearward
during gestation like it does in other apes. Further, the
Pleasure Proposal propounds that the prominent penis arose
by sexual selection during hominid evolution. Thus, males
with the longest penises were selected since they facilitated
climaxes in their partners. Plausibly, the climaxes were
emphasized with cries of pleasure while copulating. Goodall
says, while mating the female chimp gives "high-pitched
copulation screams (or squeals)."

Awkward position of human vagina

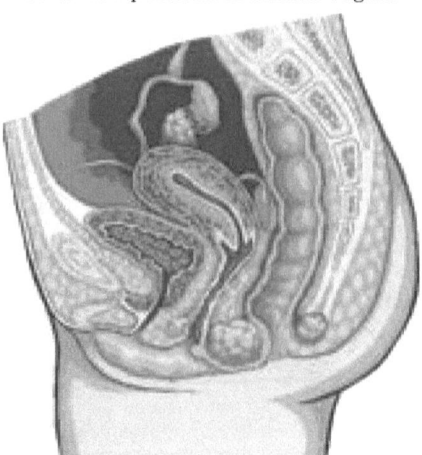

Thighs interfere with entry of penis

Guttural utters highlighting the delight of men, women and adolescents during sexual intercourse is often depicted in romantic and racy novels. Metalious writes in *Peyton Place* about a teenage girl vocally emphasizing her advance toward orgasm while copulating with her married lover. "She began to make moaning, animal sounds." Copulatory utterances aren't restricted to primates since cows and pigs moo and oink respectively while breeding. Still, human intuition questions an orgasm in less cognitive animals, especially one accompanied by sounds of pleasure. Nonetheless, the animals weren't exposed to sexual stimulants such as graphic magazines when they were young.

"Certainly," Goodall says, the home-fostered Lucy who gaped at *Playgirl* "masturbated almost daily after attaining sexual maturity, appeared to achieve orgasm; at the climax of a bout of self-stimulation she laughed, looked happy, and stopped suddenly." Goodall says when chimps anticipate copulating there is most always a male courtship display that "signals his sexual arousal and "attracts the attention of the female." When aroused males walk upright, their erect penises are easily seen, a supportive indicator that the Pleasure Proposal applies to chimpanzees whose penises are relatively longer than gorilla and orangutan penises.

Moreover, a theory holds sexual selection favoring upright males with the longest penises led to bipedalism. Additional support for the theory is a courtship display invented by a male chimp. Goodall says the display combined a bipedal stance with an erect penis and direct gaze. A requisite for the Pliocene penis to begin elongating and keep it up is bipedalism was adaptive to sustain survival and reproduction. A flasher theory of sexual selection for bipedalism "may seem frivolous," Falk says, "or even humorous. Nevertheless, it is noteworthy that relative to other primates, human males have enlarged penises."

A bonobo demonstrates a flasher theory posing
apes standing with erect penises led to bipedalism

A chimp's testes are super-adaptive according to the Sperm Competition Theory that proposes large reproductive organs are selected for species that mate frequently. Thus, for chimps, Nature picked out big testicles for abundant sperm and long penises for efficient delivery. Consequently, the male with the biggest testis and longest penis to deeply implant plentiful semen has the best chance to fertilize the egg and pass on his genes. Human testicles are super-adaptive by their role in increasing world population and sub-adaptive judged by their failure to supply oodles of sperm to pump through a big penis.

Reproductive structures are tagged super or sub-adaptive since adaptation is a concept and like any concept, may become intuitive. Natural selection and speciation are theories that are second nature when conceptualized. Conceptualizing the common ancestor of African apes and hominids is made easier by studying current species of apes. "The chimpanzee and the gorilla-are genetically close to us," Falk says, "and it is therefore instructive for us to study their distinctive lifestyles when attempting to reconstruct the behavior of early hominids." Gorillas and chimpanzees emerged from two founder groups that diverged from their ancestors several million years apart.

While the apes evolved, specific genes were added to the genotypes of transitional animals turning the first apes to diverge into gorillas and the second into chimpanzees or ancestral apes. Their adaptive zones determined their phenotypic construct and development of a social organization. Mayr says "some species of chimpanzeelike ape succeeded in establishing founder populations in the belt of tree savanna surrounding the rain forest." Food was their primary concern and their individual differences would affect how well they would adapt to the new zones.

The chimpanzees incorporated "a beneficial sociality with the basic food search." Power says. Meanwhile, they optimized the survival chances of the individual and, ultimately, the species. Five million years ago an unidentified species of chimps gave rise to bipedal apes signaling the commencement of hominid evolution.

A theoretic progenitor of Homo hominids was Homo habilis that evolved from a small brain Australopith or on a separate line from the ancestral ape. The Australopith society was presumably the same or very similar to the system of the common ancestor they shared with the great apes, including humans.

A social structure unfolds in a founder population as it adapts to the new niche. The society remains essentially the same as the precursor system if the environment is similar or adjusts to different environs. Power says a social structure based on mutual reliance is "optimal in terms of foraging strategy, in the type of natural habitat to which the common chimpanzee is adapted." Chimps living in the mutual reliance system move around freely in a pattern of fission and fusion. The freedom to come and go at their leisure operates at "subgroup, local group and larger society levels."

Power says the social structure is sustained by adaptions such as open groups of chimps that range "familiar, undefended, typically overlapping home ranges, which are local units of a larger interacting, interbreeding population." Each mobile or sedentary unit includes a charismatic male or female chimp and sometimes more than one of these confident apes. The charismatic chimp becomes the temporary leader of the unit, composed of one or more chimps. The word leader poorly defines a competent chimp since it's connotes an authoritarian role, but like the word promiscuous, it lacks better synonyms.

All mature chimpanzees, Goodall says, "that whatever their status in relation to the group as a whole, may from time to time be the highest-ranking individuals (hence the leader) of a temporary association." Power defines the relationship between charismatic chimps and other chimps in a unit as "noninterference mutualism." In the mutual reliance system there is no direct competition for food that individuals obtain from available sources. The social system is extraordinarily egalitarian and direct competition or restriction of fission and fusion or freely coming and going would tend to corrupt a society that does not include privileged classes.

Privileged by wealth

Billionaire

CHAPTER 15

Gorillas don't compete directly for food and their adaptive society is composed of groupies that depend on the protective silverback. Although gorillas are gentle apes, they are mutually dependent and will aggressively rush to the aid of a threatened individual. Their quickness to protect each other is an instinct, inherent in group affinity. Falk says in their natural habitats their daily life "seems rather placid. Sunbathing is a favorite pastime. So is eating. Life in a gorilla harem is not nearly as sexy as it sounds. In fact, it's relatively boring compared with the more lively goings-on of promiscuous chimpanzees."

A Walton like gorilla family

Placid group. Not much testosterone and estrogen.
An adaptive shortage

Gorilla placidity is a consequence of their few sex hormones like testosterone oozed from their bantamweight testicles and adrenal glands of both sexes. Gorilla copulatory performance is even less than men's poor performance, although performance applied to men is another poorly descriptive word. Despite the dearth of gorilla androgens, their endocrine system is fully mature and they are well-adapted primates in a well-adapted social organization. Albeit their social structure is tiered, the position of the patriarchal silverback is benign and protective. Respect, and not fear of the big fellow, is an adaptive property of their social structure

A Rotarian kind of ape, service before self personifies the silverback's readiness to put his life on line for his group. Succored by the silverback and cohesive society, gorillas are the only apes to adapt life on the ground where dangerous animals tarry. They have solved most problems in the struggle to survive, except the one they share with many other well-adapted species. Their peaceful nature in the bipedal world is mirrored in their Quaker cousins. They and their chimpanzee kin were habituated or tamed in their natural habits by researchers such as Dianne Fossey and Jane Goodall.

Although gorillas must be painstakingly habituated, their terrestrial lifestyle makes them easier to study than some other animals. A few other terrestrial subjects studied are midwife toads, Gila monsters, ostriches, kangaroos and elephants. Arboreal animals such as gibbons and orangutans, and the aerial fowl, eagles and buzzards, are harder to observe. Gorillas, like burnt babies fearing the fire, are cautious of people. Their misfortune is a few primates with the big brain find it easy to kill their inoffensive relatives.

Fossey's gorilla study group was difficult to approach until they realized she was not one of the murderous bipedal apes that do them harm. Fossey established rapport with her hairy subjects through patience and a relaxed composure. The gorillas eventually realized she was an unobtrusive animal and accepted her friendship. Further, gorillas don't have the testosterone excitability of chimps that are usually baited with fruits for close observation. Animals lured with food are referred to as baited or provisioned and those studied without bait are referred to as wild, and the studies, naturalistic.

The longest and most comprehensive primate study is by Jane Goodall. Louis Leakey, who Goodall calls "paleontologist-cum-anthropologist," sponsored the study at the Gombe National Park. Leakey believed studying wild chimps would provide insight into human evolution. Goodall says chimps were chosen since chimps "show many biochemical similarities to humans: the number and form of the chromosomes, the blood proteins and immune responses, the structure of the DNA. The anatomy of the chimpanzee brain is more like that of the human brain than that of any other living creature, and much of chimpanzee social behavior shows uncanny similarities to our own."

When the Gombe study began in 1960 a camp was located between what was thought to be two groups of chimpanzees that often met to mingle. Later, the researchers sorted the chimps into the northern Kasakela community and southern Kahama community. Goodall had to observe the chimps with binoculars at the outset of the study since they ran away from people. She didn't recognize individual chimps at first and was "unable to ascertain whether the chimpanzees who went to the south were those who had arrived from that direction or whether there had been an exchange of members."

One of her first impressions was the chimpanzee society consisted of temporary and constantly changing associations of individuals. Her most frequent glimpse was of small bands of four to eight chimps traveling together. Now and then she observed one or two chimps leave a small band and ramble off alone or join a different association. At other times two or three small bands came together to make a larger unit.

"Often," Goodall says, "as one group crossed the grassy ridge separating the Kasakela Valley from the fig trees in the home valley, the male chimpanzee, or chimpanzees, of the part would break into a run, sometimes moving in an upright position, sometimes dragging a fallen branch, sometimes stamping or slapping the hard earth." The running displays of the male were accompanied by loud hooting and afterward he would often swing into a tree that overlooked the valley he was about to enter. He would sit and stare silently into the valley obviously anticipating a signal from the other chimps

Goodall says if another association of chimps was in the valley feeding on figs they almost always hooted back like they were answering the hooting ape. Soon the vociferous chimpanzees would hurry over the grassy ridge and down the slope into the valley. There they joined the other hooting apes in the trees where they all feasted on figs. Any females and youngsters that joined the feeding chimps intermingled with no undue excitement. They simply climbed the trees and greeted the feeding chimps and began stuffing themselves with figs.

The foliage hid many details of their social behavior, but occasionally Goodall was treated to fascinating vistas. "I saw one female, newly arrived in a group, hurry up to a big male and hold her hand toward him. Almost regally he reached, clasped her hand in his, drew it toward him, and kissed it with his lips. I saw two adult males embrace each other in greeting. I saw youngsters having wild games through the treetops, chasing around after each other or jumping again and again, one after the other, from branch to a springy bough below.

"I watched small infants dangling happily by themselves for minutes on end putting at their toes one hand, rotating gently from side to side. Once two tiny infants pulled on opposite ends of a twig in a gentle tug–of–war. Often, during the heat of midday or after a long spell of feeding, I saw two or more adults grooming each other, carefully looking through the hair of their companions." Goodall says these gatherings in 1960 at the Festival of Figs were "excited and noisy reunions" of carefree apes from the north and south feeding and mingling peacefully together.

When they had fed and mingled for a time, they left to feed elsewhere or divided into smaller groups that went north while others went south. In 1962 a charismatic male stopped by the camp on his way to feed on an oil nut palm nearby. He was the first chimp to visit the camp and Goodall named him David Greybeard. David continued dropping by and one time the self–assured ape took a banana from the table. Bananas are scarce and coveted at Gombe and anytime the outgoing David came to call, Goodall made sure he had bananas.

David came by himself for the next five months when one day brought along an excitable male Goodall dubbed Goliath because of his superb physique. David was Goliath's good friend and if Goliath became nervous, Goodall says "David would often reach out and touch his companion's groin," which is an adaptive method to relieve tension. A nervous male Goodall named William joined the two apes later at the camp. When David was savoring his bananas, the edgy William and fidgety Goliath watched awhile from behind the trees before they partook of the gratuitous treats.

The extroverted David, excitable Goliath and uptight William typify varied characteristics Power says is an "inborn tendency known to be typical of primates, a pronounced variability of temperament." The three apes were a mobile unit in the system of noninterference mutualism wherein David was the charismatic chimp who reassured his shy chums. Later, more apes, trailed by a gang of baboons, joined the chimp trio composing Goodall's first free lunch bunch. Bananas were placed in heaps to attract the apes. The fruits are so relished that one male chimp ate sixty at one sitting.

Power says the stress of bickering over bananas during the next few years led to psychological disturbances in the Gombe chimps. "When bananas were provided daily at camp in large amounts," Goodall says, "there was a dramatic rise in the frequency and the severity of food competition as well as in the numbers of chimpanzees visiting the camp at any one time." One time a pushy baboon got into a scrape with David who sought help from his beefy buddy Goliath. They changed roles and Goliath became the temporary reliant ape to support David.

Chimpanzee male and female leaders in the egalitarian society help to maintain "the loosely structured group of individuals whose goals are primary, i.e., food and reproduction," Power says. "Individuals alternate between leader and follower roles not as group goals change, but as situational needs change." There are adaptive values in changing roles when chimps fission and fuse in subgroups of two to ten members. One is, hierarchies are prevented from forming, and another is tensions are diffused before they build up until tempers flare. Fission and fusion is a valuable principle of the egalitarian social structure.

The principles are necessary for the society to operate smoothly, and if they are altered, it may confuse the members. Comparable to chimpanzee egalitarianism is the human egalitarian system that is seen in hunter-gatherer societies that have virtually vanished. Their independent foraging way of life is all but gone in the few remaining tribes. Characteristics of the societies are; food is eaten the day it is obtained or within several days. Food isn't processed or stored and tools are simple and easily made from local material. Tribal members have an immediate and direct return for their labor.

Power says foraging societies do not incorporate a "superior-inferior ordering based on physical dominance or other sources of power such as wealth, hereditary classes, military or political office. The foraging society evolved out of nomadism, open groups, lack of territoriality, loosely defined home ranges, a wide recognition of relations, and the inheritance of behavior patterns such as tool use, drumming, dancing and bedmaking of an ape-like ancestral stock. Undisturbed chimpanzees manifest all of these phenomena." The hunter-gatherer arrangement is called the original affluent society since its members only work a few hours a day and enjoy lots of leisure time.

Materialism, exemplified by country clubs, luxury cars and big houses is disregarded in the immediate return system. Constant movement to hunt and gather minimizes the need for personal goods. Omar Khayyam expressed little need for personal goods. "A Book of Verses underneath the Bough, A Jug of Wine, a Loaf of Bread-and Thou beside me singing in the Wilderness-" The hunter-gatherer society foraging in the wilderness is surmised to be the adaptive society of Homo sapiens and extinct hominids. Anthropologists deduce the early tribes lived in peace until conditions altered the principles of their social structure.

The anthropological deduction is based on artifacts of war, such as daggers that disappeared rapidly in the archeological record beyond 10,000 years ago. Primitive warfare began during the Agricultural Revolution with fighting over land or tribes raiding the crops of their neighbors. Since then, innumerable wars have ravaged humanity, and the number of unrecorded wars is unknown. It's just a guess how many future wars will plague the world. The early egalitarians enjoyed their paradise, but lost it like the couple clothed in fig leafs and the Gombe chimps, two primate species baited with fruit.

Wars were unknown when some Pliocene apes left the trees retaining their adaptive egalitarian society. Their social structure was principled similar to gorilla society except for the big polygamous male and his monogamous females. Nature didn't select a large guardian male for any descendent species of ancestral apes. The reason is conceivably attributable to the structure of attention that is "the way attention is organized in a group, within which attention to companions takes place preferentially," Power says. The structure allows all the apes to reciprocally communicate so each one has a role in supporting their egalitarian society.

The ape society is composed of three basic personality temperaments, nervous, calm and confident. Nervous chimps are 'alerters' who sense an interruption of the status quo. Through posture, gesture, facial expressions and vocal tones they broadcast their concern to the calmer chimps who are in the majority. The calmer chimps evaluate the concern and either disregard it or pass it on to the confident chimps who decide on the response. Their interactions illustrate the structure of attention in operation. Nervous and confident chimps are remindful of television's Barney Fife and Andy Taylor, and the calmer chimps, the people of Mayberry.

The attention structure of a chimp society negates the need for a big male, otherwise in every human community there might be a man twice as big as the other fellows. However, with obesity escalating in the Western World, the big guy wouldn't look out of place. Deductively, like chimpanzees, bipedal apes of the Pliocene hunted, foraged, sought safety and nested in the trees. There is no division of labor in chimpanzee society other than the females tend the children. Thus, it's inferred the same was true for the descendent species of the Pliocene common ancestor.

The common ancestor's descendants were the Australopiths, and presuming a chimp pattern, their babies easily exited the womb and males didn't assist the mothers. The australopithecine newborn is "as advanced as a newborn chimp who knows how to hold on to its mother," Mayr says. Males not assisting with the children would discourage exclusive relationships, expressed in modern jargon innocently as 'going steady' and immodestly as 'hooking up.' The Australopiths or an unknown Pliocene precursor bequeathed the egalitarian system to Homo habilis. Until the system was corrupted, it would serve the Homo lineage.

Egalitarianism was adapted by chimpanzees and functioned throughout generations until it was distorted by the species that unwittingly distorted its own social organization. The original immediate-return system of human hunter-gatherers gave rise to, and coexisted with, the delayed-return system. Power says delayed-return foragers "hold rights over valued assets such as beehives, pit traps, fish weirs and so on, which yield a delayed return for considerable labor over time." The delayed-return system combines agriculture and herding with hunting and gathering, and though it is not complicated, it's more involved than the immediate-return system.

Delayed-return societies have familiar kinship obligations and dependencies, lineages, clans and arranged marriages. The system arose after the Agricultural Revolution and some diverged into directly competitive producer and consumer economies that eventually became stratified. As the principles of their society changed, charismatic men and women gave way to authoritarian kings and queens. They commanded thou shalt do this and shalt not do that such as thou shalt go to war and thou shalt not protest. Meanwhile, some tribes like the Kung of southern Africa retained a basic immediate return system with a few features of the delayed system.

The immediate return system was adapted in the optimum setting of tropical Africa's regions of perennial vegetation. The egalitarian system did not expand globally, supposedly because it was selected for arboreal apes and not for humans living in seasonal climates. People, and not chimps, expanded the immediate return system into some delayed return societies. The immediate and delayed return societies have no political, economic, religious or judicial systems. "Formal government would destroy the egalitarian nature of the society," Power says, "and, lacking egalitarianism, the cooperative effort would collapse." Thus, the pure and adaptive egalitarianism system is solely social.

Immediate return societies, Power says, are built on "social relationships and required behavior: roles and statuses, enforced by a few simple but effective positive and negative sanctions (rewards and punishments). The prime reason behind sanctions, both positive and negative, is maintaining the delicate equilibrium that enables a hunting and gathering band to pursue its essentially cooperative, egalitarian economy." The independent hunter-gatherers are self-assured because they govern themselves through mutual cooperation and all are awarded equal status. Clearly, some members are less assertive and skillful than others, but there are equalizing means to deter emphasis on disparities in ability among members.

Further, the society doesn't restrict people from expressing themselves through knowledge, skill or personality. Hence, "a dominance order cannot appear in conjunction with such a system," Power says. A temporary leader and the group are mutually dependent on each other's approval. Mutual approval between leaders and followers is also an adaptation of the chimpanzee immediate return system. "To maintain the basic relationship," Power says, "the charismatic apes are as dependent on less assured animals seeking their companionship as are the dependents upon obtaining it. Such mutuality of need for support and companionship operates to maintain and balance their two status/roles."

Maintaining and balancing gene dispersion among immediate return populations is sustained by mobility inherent in foraging. Saturated gene pools decrease individuality and increase genetic defects. People living in isolated areas eventually start to look alike due to stuffed gene pools. Their cultural enlightenment is stunted and they begin to think alike almost as though they were clones. Parochial life tends to become territorial as befell the Gombe chimps when they closed their community borders. On the contrary, mobile chimpanzees and tribal people maintain their friendliness and exchange genes by moving freely in and out of open populations.

Mobile foragers don't wander aimlessly, but have a fixed home range that overlaps the ranges of other open populations. Although the range is fixed, the composition of the population cannot be predicted since it fluctuates. Foragers joining another population have no special requirements other than conforming to expected behavior that is essential to maintain the system. On the down side, closed populations do not engender the expected behavior, inbreeding is encouraged and they often become territorial and hostile to outsiders. Populations in immediate and delayed return systems are divisions of an open and accessible society with cohesive family units.

Complementing the cohesive family units are kinships with a social quality as well as a bloodline origin. Social kinships unite people over great distances, giving them a sense of belonging without assigning them specific obligations. Intricate kinships in an open society would oppose equality because it would involve specific loyalties. Clearly, if people inherited a sense of social kinship from their Pliocene ancestor it would explain the puzzling incidences of altruism toward nonkin. Altruism toward nonkin is exemplified by the Good Samaritan deed such as someone jumping in a freezing lake to save a stranger.

Altruism expressed in Good Samaritan deeds is one adaption helpful for the egalitarian social organization to operate smoothly. As the egalitarian system evolved, an emotive tone was selected that Power says is best described as "love in its broadest sense. It includes such qualities as gratitude, respect, trust, admiration, goodwill, friendship and so on." The emotive tone sounds like a list of moral virtues, but the 'virtues' are essential properties that promote positive sanctions within the society. Since the egalitarian system is not materialistic, a positive sanction or reward would not be a gift retained exclusively by the individual.

Rewards that approve egalitarian human and chimpanzee behavior are intangible and hard to identify since they are emotional experiences. Rewards are feelings of companionship embodied in the positive emotive tone of egalitarianism and support in belonging to a social unit. "In the immediate-return societies of humans and chimpanzees," Power says, "the deepest attachment is to the group." Strong bonds forge group affinity, whether they are family ties or bonds of social kinship. An inherent tendency of people to form bonds is evinced by the Stockholm syndrome, wherein abductors and hostages become friends, but it can apply to other situations.

Bonds are fortified and reinforced in the egalitarian society by conforming to the norms of suitable behavior. A social norm is an idea in the minds of human and chimpanzee egalitarians that monitors their conduct in particular situations. The closer to the norms a chimpanzee or human conforms, the better that individual is liked, which inspires further desire to conform. Thus, there is a psychological reward for conforming to norms (knowing how to act), which promotes harmony in the society. Conversely, as a hierarchical system forms, norms develop to make members feel leaders have the right to control behavior.

An egalitarian society doesn't operate without some discord though most of the time it's pleasant. Power says "there is a constant redressing of the delicate balance of the egalitarian society among both species of immediate-return foragers." Sometimes a member exhibits perverse behavior that requires dissuaders, or negative sanctions to discourage bad conduct. Human foragers sanction deviants by mocking, shunning, group ostracism or banishment. In extreme cases that rarely occur they will execute the person when the group thinks the deviant is unsalvageable and there is nowhere suitable for banishment.

Ostracism in modern societies is expressed subtlety by ignoring and not speaking to a targeted individual. A researcher observed some young children that stopped playing with a bullying child. The children "avoided or excluded him from their social activities," Power says. "He was present but, without formal declaration, isolated." Another method of ostracism is blackballing to deny a college student a fraternity of sorority membership. The denial causes hurt feelings though it isn't a group sanction to correct bad behavior. Cliques in groups ostracize people, and though it may be unintentional, it's unpleasant for people outside the group.

CHAPTER 16

The egalitarian populist Carl Sandburg hated the word 'exclusive,' frequently a hoity-toity adjective for restrictive organizations. Ostracism can work in reverse such as a member high in the socioeconomic hierarchy quitting a club and leaving the less affluent feeling ostracized or shunned. Banishment in ancient Greek and Roman cultures was so traumatic that the banished person often committed suicide. Any kind of rejection is a painful variant of ostracism and is instinctive, plausibly genetically ingrained by the Baldwin effect. The severest deterrent to deviant conduct in any society is capital punishment, or permanent ostracism.

Executions are evidenced throughout recorded history for infractions as innocuous as stealing a loaf of bread to feed a hungry family. Capital punishment is an extension of egalitarian sanctions and one dreadful drawback is the condemned is not always guilty. Chimps do not execute other chimps to ostracize them eternally as do their larger brain relatives. Ostracism by chimpanzees to discipline deviants is a question scientists haven't resolved. Chimps in the wild enter and exit subgroups so often that what might be interpreted to be ostracism could be an instance of shunning or nothing except random movement.

Ostracism, by one definition, is the spontaneous result of similar individual decisions by group members that requires a certain cognitive ability. Goodall says ostracism, as a social modulator in human groups hasn't evolved in a "sophisticated way in chimpanzee society." However, Power says chimpanzees have the cognitive ability to apply and recognize "the application of, sanctions." Further, De Waal thinks chimps are fully aware of the functional effect of their actions. Provided ostracism is a Baldwinized instinct it functions easily, but preplanning an execution to emphasize conformance to norms is an involved and sometimes heartrending procedure.

Humans and chimpanzee egalitarian systems are sustained by conformance to norms and chimpanzee behavior convinces some researchers that chimps use functional sanctions. Further, Power says, free sexual choice of both sexes "makes possible ostracism from mating of individuals of either sex whose uncooperative, deviant behavior threatens the smooth functioning of the social system." Ostracism is more comprehensive than shunning since an individual may shun another without it being a group decision. Shunning is a passive means of control without physical force and can be used on stronger and more aggressive members.

Exclusion by sanction is devastating to a human forager and is inferred unpleasant for a devilish chimp. The brawny Goliath was a jittery chimp who apparently was shunned in the early days of the Gombe study. Goodall considered him a belligerent ape since at the time he was the only male she saw attacking a female. At one point Goliath was so aggressive and rambunctious that Goodall thought he was insane, though he could have been smarting from an earache or infected tooth. One time the hairy brute kicked a youngster out of his nest and took it for himself.

Ostracism as an exclusion from the opportunities to successfully reproduce thwarts the selfish gene's mission. Power says "sexual refusal ensures that the society-threatening individual's genes will not continue in the gene pool." Provided refusal is genetically motivated to insure the troublemaker's genes aren't passed on, the behavior is an adaption. Credibly, the behavior, as other behaviors, was Baldwinized in the chimpanzee genotype. However, purging a deviant's genes is pointless when the bad conduct is not heritable, but a result of something else such as redirected aggression caused by prolonged frustration.

"A powerful, repeated or accumulating frustration almost invariably results in some form of aggression," Power says, "not always or necessarily expressed directly." Goliath's aggression, for instance, was possibly redirected from an aggravating situation when he attacked the female and usurped the youngster's nest. Goliath's counterpart in the 'civilized' world is a frustrated spouse and child abuser. Clearly, keeping heads above water in a nonadaptive society is often frustrating and overwhelming as it was for the baited apes. At the start of the 1960 Gombe study the apes hugged and kissed, but sadly, gloom and doom awaited the convivial Gombe chimps.

Goodall says if the Gombe study had ended after ten years, she would think chimpanzees were far more peaceful than people. Since the study continued beyond the first decade, "we could document the division of a social group and observe the violent aggression that broke out between the newly separated factions." Power says their aggression was caused by the prolonged feeding methods that removed them "sociopsychologically from the relaxed (positive) social environment of indirect competition and put them under abnormal stress for a very long period of time, which resulted in change to a negative mind set."

Scientists who studied the Gombe publications say they chronologically reveal a radical and detrimental change in social behavior. Step by step "the whole social structure of the provisioned Gombe group is distorted and malfunctioning," Power says. The prolonged feeding system forced the chimps to obtain food by direct competition for which they were not adapted. Competing directly under irritating circumstances for withheld bananas over several years was the primary factor that corrupted their egalitarian social system. Nonetheless, Power says, there is "no firm agreement on the social organization of this species."

"Obtaining food is considered to be the most basic of physiological needs. Interference with the ability to do so can be expected to be more deeply disturbing than overcrowding, or even restricting access to mates," Power says. "In the wild, food may be scarce from time to time, but, as we know, the natural form of food competition of chimpanzees is indirect, the separate simultaneous seeking of the same resources." Goodall says the prolonged feeding had "a marked effect on the behavior of the chimps. Worst of all, the adult males were becoming increasingly aggressive." Alas, their days were numbered.

The countdown of their carefree days began in 1962 when bananas were dispensed through several feeding systems, which were unsatisfactory. Another method was installed in 1965 when banana filled concrete boxes with steel lids were sunk in the ground. Wires controlled the lids and when the fewest baboons were around a container was quickly opened like a Jack-in-the-box. The boxes were meant to alleviate the trouble, but it didn't work. When each box was opened the fighting over the coveted bananas became intense and the baited apes started to riot.

Some chimps became desperate and trimmed the end of a stick "as a 'chisel' to push into narrow openings," Goodall says. Every frantic ape acted like a child trying to pry open a cookie jar. They could have avoided the turmoil by feeding on other fruits in nearby trees. Still, they persisted and researchers reviewing the Gombe data were puzzled by the bizarre conduct. Perhaps animals are inclined to get something for nothing when it seems easy. The aroma of the hidden bananas tantalized the chimps, but the delay in opening the boxes blocked their goal of obtaining the fruits.

Blocking a motivated individual from "achieving a sought-after goal contains the constituent elements for producing frustration," Power says. Frustration produces an emotional disturbance that interferes with attention, planning, thinking and other constructive mental processes. Psychologists maintain people who can't cope with their problems are those most likely to murder someone or commit suicide. Troubles that frustrate people involve complex situations that are hard to handle even with big brains. The smaller brain chimps were just as frustrated in dealing with closed boxes as people are in dealing with complicated issues. Truly, the two primates differ in degree, not in kind.

Natural selection obviously did not adapt the chimps to deal with the artificial environment established at Gombe. Similarly, Nature didn't prepare people to deal with a nonadaptive system, so it's learning by experience through trial and error. Perhaps over a few million years evolution could adapt both primate species to deal with direct competition, a distorted principle of the egalitarian society. Another principle of the society is fission and fusion, that when violated, leads to frustration. Some caged chimps, for instance, were so frustrated they beat their heads against the bars until the poor things became insensible.

Possibly, is their desperation the chimps had premeditated suicide, although it can't be proven. De Waal tells of a death by murder most foul that struck down a male chimp confined in a Holland zoo. Since murder presupposes malice and forethought, the death was presumed premeditated by two of his cellmates. The bloody deed was conspired and inflicted after years of conflict among the three apes in a struggle for dominance and sexual privilege. Hopefully, when apes learn to communicate better, they might reveal how much they understand death and murder and suicide.

Goodall tells of another tragedy in a Florida attraction involving a "dominance struggle ending in death" when two males savagely killed their cellmate. Murders in penitentiaries are not uncommon since the frustrated prisoners, who are often psychotic, cannot fission to let tempers cool down. Similar to the frustrated chimps, frustrated prisoners reaching the boiling point can start riots. These pitiful prisoners are primates who were born with egalitarian adaptions, but can't function in a nonadaptive society. Escalating frustration and aggression aren't characteristic of undisturbed human foragers and chimpanzees. Natural selection bestowed adaptions on the egalitarian society that minimize frustrating situations.

Nature's bestowed adaptions are; the emotive tone of love and its component feelings, positive and negative sanctions to encourage conformity and discourage nonconformity, foragers move from place to place which prevents the centralization of authority, indirect competition for resources averts aggression, fission and fusion of members in mobile subgroups deters dominance orders from forming, individual self-assurance, variation of temperaments from timid to confident individuals. The adaptations are called principles and properties that support the egalitarian society. Deterioration of the egalitarian society of Gombe chimps was attributed to factors deduced by analyzing the hard-won data of independent researchers.

Power assessed the Gombe data and estimated 1965 the year negative changes began in the chimps. Prior to 1965 the chimps gave the impression they were jolly good apes. Goodall's study between June 1960 and December 1962 convinced her the "chimpanzee groups could not be considered separate communities because they freely and frequently united and mingled without any signs of aggression." Wherever wild chimpanzees have been studied,

Power says, they "are not merely nonaggressive, nonhierarchical and nonterritorial, but highly positive in all aspects of their social life.

"Because of the functional necessity of positive relations, the wild chimpanzees rarely have reason to demonstrate the negative emotions and behaviors of which they are capable, so evident in the behavior of captive primates and the recent behavior of the Gombe chimpanzees." Just as evident is the recent behavior and negative emotions of their human relatives that began a short 10,000 years ago during the early Holocene Epoch, the Age of Paradise Lost. The Beatles were prophetic. "Yesterday all my troubles seem so far away, now it looks as though they're here to stay. Oh I believe in yesterday."

All the yesterdays before the Kasakela and Kahama chimps divided researchers considered them two of several overlapping groups that made up a large population. Foraging for food by immediate return chimps is based on mobility and the principle of fission and fusion. The larger society contains subgroups of males, childless females and adolescents that forage farther from the core area. Subgroups of mothers with children and aged or infirm apes forage at slower paces. The subgroups are smaller units or microcosms wherein charismatic chimps guide the subgroup. The chimps may freely change roles or enter or exit any subgroup.

When foraging chimps happen upon a food source they hoot to broadcast its location to the group or a neighboring group. When subgroups of the same core area come together it is called a reunion, and when different core areas meet, it is a carnival. The reunions can be uneventful or exciting, and the longer the chimps are apart the livelier the reunion. Power says "like humans, chimpanzees greet others they see very frequently in a much more perfunctory manner than friends and acquaintances they meet with less often." Further, at both chimpanzee and human hunter-gatherer carnivals greetings are vigorous.

Carnivals are numerically larger than reunions and are exhilarating affairs that last longer than reunions. Scientists view carnivals as adaptive to promote socialization among populations of immediate-return foragers. When chimps first begin to assemble for a carnival their displays are tempestuous and spectacular which helps to relieve tension in the excitable apes. Goodall says the uproarious displays have a positive tone expressing "merely excitement and pleasure." After a while the chimps settle down to the calmer social activities of playing, copulating, grooming and generally monkeying around.

Power says chimpanzees and human foragers "extend fellow-feeling, or in anthropological terms, social kinship, to all of their species with whom they come into contact occasionally, yet fairly regularly." Wild chimpanzees at carnivals make new friends and establish rapport with prospective sexual partners. Other adaptive aspects of carnivals are reinforcing bonds of group affinity and exchanging genes among populations through emigration and migration. Carnivals of human foragers last a few days to a few weeks and usually convene at an ample source of food and water. However, the abundance of food is a secondary reason for the jovial occasion.

Carnivals are periods of intense social interactions with activities such as visiting, feasting, gambling, hxaro, "marriage brokerage and trance dancing," Power says. Hxaro is between designated partners from each tribe who exchange gifts whose value is unimportant. Hxaro is difficult to comprehend in Western society since the value of most gifts is important. Hxaro in human foraging systems is a cultural adaption to foster good relationships among the tribes for social and practical reasons. One reason is when a tribe's environment is exhausted and needs time to recover, the tribe breaks up and the members join another tribe.

When the land is eventually restored it may be repopulated by its former inhabitants and may become a refuge for other foragers vacating weakening areas. Making friends among the tribes before an environment falters facilitates the integration of new, but familiar foragers. Friendships are rejuvenated during chimpanzee and hunter-gatherer carnivals that are comparable to group picnics, church suppers and square dances. Power says studies show that successful businesses "parallel the chimpanzee and forages' carnivals through frequent celebrations such as Friday afternoon beer busts, sporting activities and so on, which bring many employees together in an enjoyable, informal atmosphere."

The basic social organization found in successful businesses may not be a product of industrial society Power says, but "based on fundamental social forces that human beings share with primates." Nonhuman primates aren't at the complicated level to share the social force of marriage brokerage at human-gatherer carnivals. A type of marriage brokerage exists in modern society that are often rituals based on socioeconomic status. The only analogy and speculative precursor of marriage observed at Gombe is a consortship, which Goodall calls a 'honeymoon.' A honeymoon is when the 'groom' whisks his 'bride' away for an amorous interlude.

Dearly Beloved

Consorting Chimpanzees

Aba-daba-daba honeymoon

A chimpanzee 'marriage' is followed by the honeymoon when the male whisks his 'bride' away for an amorous interlude. Goodall says if a female is reluctant to consort with a demanding male he "may display toward and all around her—he may also attack her." In contrast, Power says, "the evolved chimpanzee pattern of mating is free choice, based on indirect competition and the freedom of either sex to request, accept or decline mating. The possessive and consort forms are largely post-1965 innovations at Gombe," and are more like shotgun weddings than romantic interludes.

Observers of wild chimpanzees never saw frenzied courtships, possessiveness or attacks on estrus females. Wild chimps are sexually uninhibited and both genders express pleasure during sexual intercourse. Sexual coercion by threats or violence is called rape in Western society and date rape is sex obtained by trickery. Most rapes are driven by a need for power rather than sex. Ape rape at Gombe is contrary to the endearing behavior of the mild-mannered male at the 1960 Fig Festival who took a female's hand in his and kissed it with his lips.

Did his love turn to

hate in the Primal War?

When a gentleman, like the mild-mannered chimp, kisses his betrothed and they marry, they enter an institution with a baffling history. Relatively recent, same sex marriages have become popular despite puritanical objections. Marriages can be elaborate or simple like an ancient Egyptian ceremony when the couple broke a jar between them. Increasing divorces, extramarital liaisons and marriage counselors are equated with liberalizing social mores in the mid 20th century. Moreover, polygamy is the prevailing marriage arrangement. Universal polygamy suggests an exclusive union is a cultural tradition and not adaptive like the monogamous pairing of gibbons and turtledoves.

The affluent circles of the Western World are noted for marriages that are costly extravaganzas. Regardless of grandness, the permanence of a marriage contract is frequently violated. Most Western marriages have religious vows to forsake all others and remain together until a partner dies. Whoever performs the marriage usually tells the blissful couple their union will minimize life's tribulations. The ceremony is upbeat and often followed by an exciting reception, but it's not mentioned that half of marriages end in divorce in five years and many unhappy marriages continue because of children.

Conceivably, seeds of marriage were sowed when babies became increasingly dependent and mothers needed help. Confining relationships might have developed between hominid couples mutually caring for children, shades of the Stockholm syndrome. The exclusive relationships became a cultural tradition to help babies survive and reach reproductive maturity. Infant dependence increased as testicles shrunk in each new Homo species. Puny testicles in a polygamous species would limit a fervent male's copulations and number of offspring. The most helpless babies and smallest testicles wound up in humans, yet the polygamous instinct for frequent pleasures remains an insatiable vestige in their minds.

Relative sizes of chimpanzee and human testicles.
Big size difference is contrary to DNA similarity

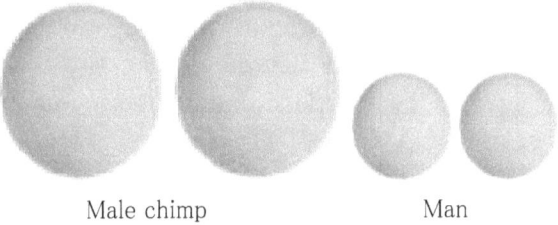

Male chimp   Man

Maladaptive structures, like puny testicles, can strain a species against the grain of its genetic directives. Straining denotes the species is grappling with elements of precursor genotypes, though it doesn't affect survival. The undisturbed male chimp who regally kissed the hand of a female suggests a tolerance for exclusive relationships. Nonetheless, an exclusive chimp union isn't adaptive and the huge testicles would make the union more challenging than in Western marriages. Still, chimps are a discriminating and sentient species whose males and females temporarily pair up with preferred sexual partners.

One instance of partner bias was noted when two females were sitting together and staring at the penises of two males. When the males took action they copulated more with one female and with more gusto than with the other one. Goodall says the favored female had a better personality that was likely cultivated during her contented childhood. The two females responded quickly to the males, but the less favored female tended to be "tense when she crouch-presented." Perhaps her tension resulted from a stressed nonadaptive childhood, or some females become anxious when observers watch them having intercourse.

Tension is often experienced in the chimp's sister species by virginal brides during the consummation of marriage in a moralistic society. Many married women raised in the prudish atmosphere of straight-laced societies have never achieved an orgasm despite extensive therapy. They are usually older women since virginity is a rarity in Western society. Metalious writes in *Peyton Place* set in early 1940s Victorian New England that the experienced married man who brought the embarrassed and virginal teenage girl to an organism was "practiced and polished, an expert who regarded the making of love as a creative art."

Many newly married men in those priggish settings were unpracticed and unpolished with no sexual experiences. They needed a seasoned woman to help them make it through at least one night. Since *Peyton Place* was published in 1956, much of the puritan influence in the Western World has dissipated through changing social norms. Victorian morals have been replaced by the 'new morality' that was fully swinging during the latter 20th century. The new morality or lack of, attest that Metalious's description of the "practiced and polished" lover should apply equally to modern adolescent girls.

Statistics indicate most of the girls are practiced and polished and it's credible their sexual voracity is inherited from a precursor's genotype. At Gombe, "adolescent females are sometimes insatiable," Goodall says, "and repeatedly solicit males of all ages, even going so far as to tweak flabby penises." Clearly, the fervor of a typical teenage girl was mirrored by one of her primate cousins. "A young female backed onto the erect penis of a feeding male," Goodall says, "and herself achieved intromission." Certainly, the sexual pleasure of youth is a wonderful thing that isn't shamefully wasted on young primates.

"Indeed," Goodall says, "an ardent old male once leaped into the nest of a young female and courted her vigorously before she had even sat up." A bipedal analogy is an ardent old man crawling in bed with his young girlfriend before she can put on her lipstick. Love in the afternoon for the old gent and chimp wasn't as ardent because "copulatory rates tend to peak early in the morning." Wild chimps don't have abstinence periods to accumulate hormones that fire their passions since they have sex galore like their free loving cousins of the Western World.

CHAPTER 17

Amoral sex is without the stress inherent in directly competitive societies. Power says sexual behavior of wild groups was "remarkably relaxed, non-competitive and amiable."

Copulation between wild chimps is causal and uninhibited as it is with free loving people who haven't been brainwashed in puritanical settings. People and chimps have commonalities that far exceed the distinctions that classify the two species. "The differences between chimpanzees and humans are not trifling, but they do not arise from any difference in kind," Gould says. "Part by part, order by order, we are the same; only relative sizes and rates of growth differ."

One part that's the same is male chimps and men like older and experienced sexual partners. "Older women make the best lovers," lyrics fitting for chimps. Goodall says some ardent suitors following an older estrus female "braved the novelty of our camp in order to keep close to the object of their desire." Another instance of picking riper fruits was two tempting females were near a lusty male, and thirty out of thirty-eight times, "the older of the two was chosen." Some older men and women enjoy partners their grandchildren's age, credibly a psychological need inherent in a noadaptive society.

On the other hand, the amorous appeal of older female chimps might be adaptive to sustain the birth rate since they reproduce throughout old age. More evidence the appeal is adaptive is older and more experienced females "typically wait for additional signals to clarify the male's sexual motivation," Goodall says. Considering the feelings of a sexual partner could be an adaption of both humans and chimpanzees exemplified by the "practiced and polished" lover of *Peyton Place.* The lover perceived the temerity of the young virgin and, calling on experience, prolonged his foreplay in guiding her through the preliminaries of sex.

Thoughtfulness is an emotive affiliate of compassion that may increase the number of offspring reaching reproductive maturity. Offspring sustain the continuity of a species, and the more impregnations, the more offspring. Fertilization is facilitated in women when they are physically and emotionally relaxed, which is inferred the same for female chimps. Altruism expressed as empathy engenders a pleasant and relaxing intimacy that enhances the desirability of a male or female as a sexual partner. Still, as attentive as wild male chimps are in their natural habitats, their plentiful testosterone would hamper a lasting relationship.

Relationships between two wild chimps are temporary since they fission and fuse to diffuse tensions that would develop in a lengthy union. Certainly, if they don't part they can't meet. People regularly parting and meeting would relieve the tensions arising in protracted unions, like marriage. The principle of fission and fusion along with indirect competition operate to prevent social and personal conflict. Although polygamy isn't an egalitarian principle, but a chimpanzee reproductive strategy, its 'free love' aspect discourages jealousy inherent in exclusive relationships. Possessiveness, jealousy, anger, hatred and the other negative emotions fester in a nonadaptive social organization.

Possessiveness raised its ugly head at Gombe when baiting the chimps disrupted the principle of fission and fusion. Direct competition for bananas brought on an array of behavioral changes that profoundly affected their adaptive social structure. Goodall says "the disadvantages of the feeding system from 1964 to 1967 became more and more apparent. Ranging and grouping patterns, feeding, and aggression were increasingly influenced. The quality of the

relationships of these and other chimps was undoubtedly
affected by this more-frequent association." The chimps were
baited at the camp located on the edge of the ranges of some
chimps.

The chimps congregated at the camp and slouched around
waiting to be fed like loafers wasting time at a beer joint.
However, they didn't have the casual disposition of loafers
since the withheld bananas blocking their wish goal frustrated
them. When undisturbed chimps feed in trees they spread out
and peacefully pick fruits. Picking individually on adjacent
limbs exemplifies indirect competition, a principle of the
egalitarian social organization. Swiss philosopher Jean
Rousseau in 1762 proposed a system wherein no one held
authority. Rousseau envisioned an egalitarianism society
wherein everyone had equal rights. Clearly, he was barking
up a slippery tree.

Altruistic people like Rousseau have proposed classless
societies throughout history. Egalitarianism and altruism are
embodied in the Golden Rule advising people to treat one
another like they want to be treated. The Golden Rule is a
moral precept of most major religions and credibly an
adaption ingrained in the genotype of humans and chimps.
Nonetheless, the Golden Rule epitomizing egalitarianism and
altruism can't function optimally in a nonadaptive society.
Egalitarianism is a system wherein everyone benefits equally,
but it doesn't include hierarchal structures innate in
capitalism wherein people and nations compete directly for
resources and territory.

Despite the fierce competition of capitalistic societies,
egalitarianism and altruism appear qualities people adapted
long ago. Throughout recorded history the qualities
functioned to some degree even in despotic societies.
Capitalism was called political economy by Scottish
philosopher Adam Smith who justified the system in the late

1700s at the peak of the Industrial Revolution. Smith said the profit incentive should be left alone because an "invisible hand" would operate and benefit everyone, especially poor people. Sixty years later German economist Karl Marx berated capitalism in his *Communist Manifesto* because of inequality inherent in capitalism or free trade.

Marx wanted capitalism to be overthrown by violence if necessary and replaced by communism, an idealistic system of equality. "I laugh at the so-called practical men and their wisdom," Marx says. "If one wants to be an ox one can easily turn one's back on human suffering and look after one's own skin." Seventy years after Marx's publication, Russia overthrew its monarchy for communism that persisted for seventy more years until it collapsed. Attempts at communism before Marx were unsuccessful and its continuance in current communist systems that practice free trade is uncertain.

Egalitarian and classless communism is principled by indirect competition and unrestricted movement, which can't be maintained in a directly competitive economic system. A principle of capitalism is direct competition, and that alone is enough to keep the reestablishment of the human adaptive society an unfilled hope of Rousseau and Marx. Adam Smith defended capitalism in *The Wealth of Nations*, which he might have forgone could he have seen the consequences of unregulated capitalism a few decades later. Unchecked capitalism leads to monopolies controlled by rich and politically powerful people, blatantly obvious in the Western World.

A banana monopoly controlled by powerful chimps emerged in the early days of the Gombe study. The monopoly was seeded by direct competition over the fruity treasure buried in closed boxes. The boxes, like hidden treasure chests, that were installed in 1965 ended the free access the chimps had to the booty. Increasing their frustration was a horde of other chimps and baboons nervously waiting for the inadequate supply of bananas. The frustrated chimps reacted with aggression, but not toward the boxes. They attacked each other and fought like trade unions and big business corporations or nations against nations.

The nonaggressive Gombe society deteriorated into a microcosm of a capitalistic society. "Relations between individuals competing directly for the same objects tend to take on a negative tone and may become openly unfriendly or hostile," Power says. "Since more time is devoted to self, others are assisted less. Much of the positive behavior that supported the peaceful, successful social order changed radically and negatively, or disappeared, to be replaced by socially destructive behavior." The gathering storm built steadily from 1965 until the winds of war swept through the northern and southern communities of Gombe chimps.

The northern chimps so outnumbered the southerners that the conclusion of the Primal War was as predictable as the American Civil War. "Has any one of you gentlemen ever thought that there's not a cannon factory south of the Mason-Dixon Line?" Rhett Butler said. "Or how few iron foundries there are in the south? Or woolen mills or cotton factories or tanneries? Have you thought that we would not have a single warship and that the Yankee fleet could bottle up our harbors in a week, so that we couldn't sell our cotton abroad?

"The thousands of immigrants who'd be glad to fight for the Yankees for food and a few dollars, the factories, the foundries, the shipyards, iron and coal mines-all the things we haven't got."

The Primal War like the Civil War and other horrifying wars of their bipedal kin, was replete with broken bodies and suffering. Goodall says the events that led to the gruesome deaths of the southerners "are, I believe, unique in the history of primate field research." Truly, none of the independent observers had ever seen anything like this turmoil among wild chimpanzees.

When Goodall observed these same chimps in 1960 at the peaceful Festival of Figs she saw some arriving from the south and others from the north. She felt they were two social groups that often met and were part of a larger population assembling for a carnival. Their gatherings indicated the chimps were overlapping segments of a fluid social system. In 1966 the Gombe researchers concluded some chimps spent more time in the north and others in the south. The researchers labeled them the northern and southern subgroups.

The social structure of the northern and southern chimps went through a qualitative change that led to the Primal War. Power says unintentionally the protracted artificial feeding at Gombe "deeply frustrated the chimpanzees which precipitated extensive, qualitative change in their behavior and organization." Paralleling the Gombe chimps in the bipedal world, the events that led to the American Civil War, or virtually any war, began during the Agricultural Revolution. The societal changes at Gombe were unintentionally forced on the chimps and occurred 10,000 years after people unintentionally corrupted their own system.

The relics of warfare fading out before the Agricultural Revolution date the negative change of the human adaptive society. In theory, to compare prehistoric and modern society, the nearly vanished hunter–gatherer society is contrasted to current society. Similarly, to compare the undisturbed and disturbed societies of Gombe chimps requires two sets of evidence. The Gombe researchers report one set of evidence and naturalistic observers report the other set. Analyzing and comparing both sets of evidence indicates the potential of chimpanzees. Thoroughly analyzing human potential would require more evidence than simply the lack of artifacts of war before the Agricultural Revolution.

Since more evidence is needed, the evidence is taken from studies of human hunter-gatherer egalitarian systems and modern society. The studies infer the structure of the human adaptive social organization and that the hunter-gatherer system didn't change significantly until the arbitrary date of the mid 20th century. The adaptive social organization of social primates such as humans and chimpanzees evolved over many years. Adaptations that sustain the organization were optimized by natural selection so the society resists modification. Moreover, social animals are endowed with behavioral adaptations that are helpful in dealing with adverse situations.

When adverse situations occur that threaten the survival of a social animal, latent behaviors may surface and destroy its social organization. Indeed, the behaviors are adaptive if the society needs to be demolished for the population to survive. Power says suspending or relaxing the positive behaviors that support the mutually reliant system, and switching to a negative set that breaks the system down may be "a prerequisite to a rapid change to another form of social system." The usual crisis that can affect the social structure of social animals is a severe and protracted food shortage.

The crisis can result from natural calamities such as prolonged droughts or floods that reduce nutritional resources. Food is the primary necessity for a species to survive, and lack of it demands drastic measures. Starvation is a powerful impetus that can drive a prey animal to bite its predator. The deadly measure was demonstrated in a laboratory experiment when a famished mouse bit a cat that ate the mouse. During a famine "the open system might become disadvantageous," Power says, "and rapid social change to a different form of organization might, in crisis circumstances, be a survival mechanism."

Curiously, no food crisis existed when the Gombe chimps converted from a free and open social system to a hierarchal structure. True, there's some food scarcity during dry season, but nothing like a natural disaster. Chimps manage the dry season by their adaptive method of fissioning when they feed separately and silently without food-calls. There were sufficient resources in the forest when the chimps converted to a system supported by principles opposite those of their adaptive open system. A few years after the introduction of baiting, they adopted despotism, a territorial system with defense of resources.

"Under varying conditions," Montagu says, "and especially conditions which humans have introduced into their lives, chimpanzees will undergo profound changes in their social lives, very much resembling the kinds of changes which humans have undergone in the course of social evolution." The Gombe chimps have commonalities with the Roman Empire that began with Augustus Caesar's reign 2,000 years ago. The Romans guarded and extended their borders by killing or conquering opponents, whichever suited their needs. In 300 A.D., the Empire encompassed most of the known world, but deteriorated internally until it collapsed at the arbitrary date of 436 AD.

Meantime, the selfish aristocrats were preoccupied with luxurious pursuits, especially sex. Their sexual excesses had similar components to the sexual activities of the Gombe despots. The sexual excesses of the Romans and chimps are mirrored in the Western World by commercial exploitation of the innately polygamous masses. Despotic Caesars such as Nero and Caligula intermittently ruled ancient Rome. These and other despots were often murdered with daggers as savagely as canine teeth killed the Holland zoo chimps. As Goodall and Gould say, the difference between people and chimps is not in kind, but only in degree.

The open society at Gombe was replaced by a new social order characterized by despotism. Corruption of their social organization wasn't sudden, but gradual over several years like it was guided by an invisible hand. The new order is analogous to a human oligarchy wherein one or a few rulers control the populace. The social order of ancient Italy was controlled by Roman Caesars and the Senate backed by the Praetorian Guard. Other rulers and oligarchies were England's Henry VIII and Renaissance Italy's Medici family respectively. Ancient Egyptian hierarchies headed by a Pharaoh and administered by priests exemplify theocratic autocracies.

Theocratic hierarchies arose from loosely knit echelons of superstitious elders and their tribes after the Agricultural Revolution. Governmental, military and systems of social arrangements evolved into the stratified order of modern systems. An example is the republic type of government controlling the United States wherein laws and decisions are made by the people's representatives, but not necessarily on a plurality of their opinions.

The major influence in America's economic policy is big business that critics call a waxing oligarchy. The invisible hand that moved rapidly after Adam Smith vindicated capitalism drives free trade practiced by directly competitive nations.

Proponents of regulating American capitalism argue commercial interests control the federal government's three branches. Further, preferences of a dominating minority are increasingly supplanting the needs of a subordinate majority. The invisible hand that is widening the gap between the haves and have-nots guides global economics. "In the earlier epochs of history we find almost everywhere a complicated arrangement of society into various orders, a manifold gradation of social rank," Marx says. "In ancient Rome we have patricians, knights, plebeians, slaves; in the Middle Ages, feudal lords, vassals, guild–masters, journeymen, apprentices, serfs; in almost all of these classes, again, subordinate gradations."

Conversely, in human egalitarian societies there are no subordinate gradations because there are no breeding grounds for political office, wealth, hereditary classes, and military units. These features arose from the corruption of the hunter–gatherer adaptive societies, which led to directly competitive economics. What's more, hierarchal systems of the Western World with economics at the foundation are organized and managed from the upper echelons. Some economists prognosticate that total world globalization will be in operation by the year 2050. Provided that materializes before a nuclear Armageddon, it forecasts an oligarchy will regulate ten billion people.

The event of a global oligarchy would suppress and assign people to natural selection's mindless role, which is determination to survive and reproduce. A world order divided into an upper and lower class would be the ultimate corruption of the egalitarian society Nature forged for some primeval primates. Egalitarianism was ingrained in their genotypes plausibly by the Baldwin effect and bequeathed to their descendants, including humans and chimpanzees. When the Gombe chimps began changing, the hugging and kissing during the 1960 Festival of Figs was exchanged for slapping and kicking in the Kasakela and Kahama communities.

Discord intensified within and between the communities until it culminated in the Primal War, replete with violent deaths and suffering. Still, credit is due to the positive emotions of love, trust, compassion and others, which might be subdued, but never quit. The positive emotions are essential properties that natural selection adapted to sustain the egalitarian system. During multitudinous millennia normalizing selection strengthened the emotions to fortify bonds among members of the immediate-return system. Opposing the positive emotions are the negative emotions of hate, jealously, anger, vengeance and others that operate to weaken and break egalitarian bonds.

The two sets of emotions function in chimpanzees and humans as behavioral responses to varying circumstances. Anger, for example, is a response to frustration that can induce aggression. "Aggression, particularly in its more extreme form is vivid and attention catching, and it is easy to get the impression that chimpanzees are more aggressive than they really are," Goodall says. "In actuality, peaceful interactions are far more frequent than aggressive ones; mild threatening gestures are more common than vigorous ones." Similarly, in the bipedal world community violence and wars give the impression people are more aggressive than they are in reality.

Among people there are many more peaceful interactions than aggressive ones and one reason is aggressive situations are so unpleasant they are avoided if possible. Sensational incidents of extreme aggression are media favorites since humdrum news has poor audience appeal. Historically, America has experienced more peaceful periods than major wars that have occurred about every twenty one years since 1776. The longest was the Vietnam War unless the perpetual war on drugs counts as a war. When the drug war started is speculative, but like other wars is certainly bears political and economic gain for special interests.

War became functional in the emerging nonadaptive social system during the Agricultural Revolution. People within current nonadaptive systems vie for resources and mediums of exchange. Paper money, coins and sometimes gold and silver are value symbols paid for commodities and services. Nations vying internally may lead to revolutions and vying externally may initiate offensive or defensive wars. Wars are mostly economic variations of the same theme different only in name such as religious wars. Wars of expansion are essentially economic wars that often result by overpopulation sapping resources and living space. Every war exacerbates the misfortune of human existence.

Theoretical economists propose expanding global populations and depleted resources could end in a nuclear holocaust. The catastrophe is a logical conclusion to Malthus's Theory that poses food supply expands arithmetically as geometric population increase causes starvation. The theory suggests holocausts are adaptive to cleanse away superfluous people in unsustainable populations. The nuclear war on the *Planet of the Apes* had a cleansing action that precipitated a radical social change. Apes using people for guinea pigs in medical experiments instead of people using apes was a complete social change.

When the Gombe chimpanzees first began to change socially, uneasiness crept within and between the northern and southern neighbors. Goodall says before the chimps were baited the "quarrels which provoked aggressive interaction were momentary squabbles and the longest was 1½ minutes." The aggressive interactions in the early days are evidence the chimps were vexed by "human interference at a much earlier date than my chosen watershed date (1965)," Power says. The core areas of the Kasakela and Kahama chimps were near the border of the park and the number of people continued to increase.

CHAPTER 18

When the Gombe study began in 1960 the human settlements around the park restricted the distance the chimps could employ their adaptive mode of fissioning. Some of the people encircling the park walked softly and carried big spears or arrows. They carried rifles and nets or whatever equipment needed to kill or capture their fellow primates. The soft walking bipeds were a source of apprehension long before the Gombe project. Adding to their stress, the Gombe researchers set up camp between the communities near the park's border. The chimps were baited and observers followed them on foot, which had consequences.

"Certainly the presence of humans and the ongoing research, and the way the research is conducted mean that Gombe chimpanzees are not living in an undisturbed environment," Goodall says. Shadowed by curious creatures with unorthodox bodies that walked erect stressed the apes and was "surely an inexplicable behavior by an alien species," Power says. One consequence was uneasy males started challenging each other for dominance. Goodall says dominance concepts weren't important in 1965 "because of the temporary nature of chimpanzee groups, the loose form of social structure, and because aggressive and submissive interactions between individuals were at that time, infrequent."

When the males started challenging they displayed vigorously with hoots and screams. The recurrent spectacles of the apes that achieved dominance are mostly bluffs sufficient to subdue the other males. What the apes are thinking when they throw tantrums is hypothetical, but their actions are similar to schoolyard bullying. Challenges sometimes erupt into fights when the hierarchy is unstable and the chimps are excited. However, males restrict the use of their deadly canine teeth. Goodall says there are no cases of "fights (or any kind of fights within a community) that have resulted in the death of the loser."

Most males have a personal style of challenge for dominance they attain through cagey bluffs or intimidation by being large. A less energetic and shrewder strategy is a coalition of males. An ambitious male may form a coalition with lesser-ranked males to obtain the crown where under his head will lie uneasy. The guileful ape manipulates his subordinates like a crafty person climbing the corporate ladder. The manipulative ape pacifies the potential conspirators by grooming them or not interfering when they copulate. Acquiring and sustaining the shaky status of alpha ape takes time, energy, scheming and sometimes brute force.

Mike was a bumptious ape aspiring for the post of Commander-in-Chief of Chimps, who commandeered empty kerosene cans from the camp. He clamorously tumbled the weapons of mass dissonance ahead of him as he charged into a startled bunch of grooming males. "After we removed the cans," Goodall says, "Mike made determined efforts to secure other human artifacts to enhance his displays-chairs, tables, boxes, tripods, anything that was available." Since he was a smaller ape without weapons, he seized tree fronds and brandished them to bully the other apes.

Schoolyard bully

Mike commandeers kerosene cans

He struck terror into his fellows

Curiously, some males aren't ambitious for dominance although they might be big enough to go to the top. One big male, an affable ape named Jomeo tried his hand at displays and was as clumsy as a country bumpkin. Once he tried to uproot a small tree to brandish it, but slipped on an exposed root and fell on his behind. After several more awkward trials at being a big shot he gave up and threw in the towel. Jomeo wouldn't be a ladies man in the bipedal world since other suitors usually shot him out of the saddle.

Goodall followed Jomeo one day and saw that "no less than five adolescent males trailed peacefully in his wake." Clearly, in the bipedal world Jomeo could be a capable scout master, high school guidance counselor, male nurse or a postal employee. Despite his gentle disposition of a Quaker pacifist, Jomeo was a brutal assailant in the Primal War. Goodall says he probably was "responsible for many of the worst wounds, including Sniff's broken leg." Jomeo's good nature and rapport with his younger friends contrast to his belligerent and courageous action in the Primal War.

Jomeo's Quaker disposition and heroism equate him with two pacifists and Medal of Honor recipients; Alvin York in

World War I and Audie Murphy in World War II. These
heroes and chimps like Jomeo distinguished themselves
conspicuously by gallantry and intrepidity at the risk of life
above and beyond the call of duty while engaged in action
against the enemy. Another distinguished chimp, but far
below the call of duty was Humphrey, a despot of the
northern community. Humphrey enlisted to fight in the Primal
War, but instead of fighting for the cause, whatever it was, he
was a coward.

Humphrey never had any badge of courage, red or
otherwise, nor did he fight like a brave soldier in the 1864
siege of Atlanta by northern troops. Humphrey in a bipedal
war would likely be cited for being a coward and dishonorably
discharged. Although Humphrey was big and strong,
throughout his tour of duty he often avoided patrols that
went south. Goodall says the big ape's "reluctance-fear,
even to travel in the south dates from 1971." Researches
speculate he feared a coalition of two apes that had gone with
the southern Kahama community when it seceded from the
Gombe union.

Fortune smiled on Humphrey when the two apes he dreaded
went south to fight with the Kahama chimpanzees. Humphrey
might not have become the authoritarian ape of the Kasakela
northerners if the two males he feared hadn't pledged their
allegiance to the southern cause. In any case, the war showed
the Gombe bully who intimidated his fellows in the safety of
his home turf had a yellow streak down his back. Humphrey
would have been drummed out of the Union Army were he a
bipedal blue coat in one of the battles of Bull Run.

Humphrey

Community bully and battleground coward

Humphrey or any challenging ape expends much energy attaining alpha position. Once there, he expends more energy defending his station from other challengers. In the long run, he 'rules' like a king with no subjects. Goodall says when food is scarce chimps fragment and "the alpha male will have no greater advantage than the top-ranked individual in each of the small groups. In competition for meat, the older males are often more successful than those who have highest rank." Ironically, the alpha must contend with his 'subjects' without assurance of the prime cut from a captured monkey.

The alpha ape must also contend with other males for the prime pleasures of another flesh. Goodall says "only a secure alpha male can assert his mating rights in a group situation." Still, his assertions no more insure his prime pleasures than his prime cuts of monkey meat. Low ranked and crippled males skilled in sexual pursuits often copulate more than the alpha and pass on their genes. A bipedal parallel is the paramour in D.H Lawrence's novel Lady Chatterley's Lover. The lover is a charismatic man who is having an affair with his wealthy, but impotent employer's wife.

The man was a gamekeeper, and as many spirited people trapped by circumstances, was relegated to a lowly level of society. Lady Chatterley's husband was in a similar sexual boat as an alpha male having trouble with a lower ranked rival over feminine pleasure. During the confusion Goodall says, "a second male may take the opportunity to copulate with the female." When the alpha turns his attention to the gratified female, another menial ape will mate with an unguarded female. Additional intrigue; a lowly male will contrive a rendezvous with a willing partner who likes him and disdains the alpha.

During clandestine copulations females "contribute to the deception by inhibiting the copulation scream or squeal," Goodall says. The Gombe female cooperates with her sexual partner in the deception as effectively as her perfumed and adulterous cousin. Lady Chatterley kept her husband in the dark and made secret arrangements to meet the gamekeeper for their collusive affair. Equally enthusiastic for an interlude of gratification was the virginal teenage girl in *Peyton Place*. When her soon to be lover said he was going off for a weekend alone she said, "Take me with you."

Surreptitious rendezvouses have existed since autocratic societies demanded exclusive relationships. A literary work dealing with surreptitious rendezvouses is Gustave Flaubert's renowned 1857 novel *Madam Bovary*. The protagonist, Emma Bovary, became disenchanted with her dull husband and spiced her life with adulterous affairs. Another kind of rendezvous is a woman and her 'boytoy,' an energetic younger man with no allegiance to any woman. Some affluent women of the Western World maintain apartments solely for sex with their youthful swains. Sometimes the women support their subordinate boytoys who have no more ambition than subordinate Jomeo did to be alpha ape.

Subordinate males thoroughly enjoy sexual pleasures without the distraction dominant males have in controlling their partners. Perhaps subordinates have more fun because they "opt out of the adult male power struggle," Goodall says. "Evered is a perfect example of a male who, having clearly lost in his bid for top rank, devoted himself with considerable success to increasing his reproductive output." Evered and Lady Chatterly's lover were both in nonadaptive societies, and though Evered never reached alpha status, he had fewer frustrations. The gamekeeper was not only confined in a lower social echelon, he was restrained in other ranked systems.

Evered at one time was as ambitious as Humphrey, but his luck was limited to affairs of the heart. Unlike lucky Humphrey, whose fearful coalition of two apes went south with the southern forces, Evered got into a beastly conflict with two tough brothers. He was badly defeated and afterward decided to stop wasting time struggling with roughnecks. Instead, he pursued his basic instinct romancing the fairer sex away from the chaotic community. Although Evered retired early from the challenging business, his retirement was fulfilled with sensual satisfaction while Jomeo was often shot out of the sexual saddle.

Goodall says perhaps chimps run risks to attain high rank "for the psychological benefits as well." Evered's carnal gratifications credibly afforded him a psychological confidence denied him by not attaining alpha status. The hairy Casanova took an alternate course in life similar to Edward VIII, King of England who abdicated his throne in 1936. King Edward said he couldn't perform his duties "without the help and support of the woman I love." The woman was a divorced commoner and unsuited for a consortship with a British monarch like the consortship of a Gombe female with an alpha male.

Following Edward's abdication, and with the help of his lady friend, and especially the support of the Royal treasury, the abdicated king lived the rest of his life as carefree as Evered copulating at a carnival. Evered often went off alone like the lover in *Peyton Place*, perhaps because "he was consorting with unhabituated females," Goodall says. Evered's alter egos are in people who want to play the field and opt out of competition as soon as possible to enjoy life. These noncompetitive male chimps and men apparently think the price of dominance is more than they want to pay.

The price of glory is pricy for alpha males whereas females pay a pittance to play in penny-ante dominance games. "Although there is a good deal of aggressive interaction among the females, they cannot be ranked in a clear-cut dominance hierarchy," Goodall says. When lower ranked females acknowledge higher ranked females they broadcast their lesser rank with vocal pant grunts. The grunts are symbolic like women tennis players curtsying to British royalty at Wimbledon. DeWaal calls pant grunts "greetings" among zoo females and believes the acceptance of dominance is likely more important than proving the point.

The position of alpha ape may be adaptive to compete successfully for food in parts of Africa where the environment is harsh, but "there are not yet enough data to be certain," Goodall says. Naturalistic researchers haven't reported dominance hierarchies in wild chimpanzee groups. Some researchers think a hierarchy is functional in a rare, but adaptive social process. It's also thought a hierarchy could sometimes be a distortion of the process in a nonadaptive situation. Power says processes supporting the loosely structured chimpanzee egalitarian system can change when necessary "to a rigidly structured dominance hierarchy."

Provided converting to a short-term dominance hierarchy in a natural crisis is adaptive, suggests the group initiated the change. During the short-term hierarchy, the basic structure of the society remains intact until the crisis is over and then the society returns to its pristine state. Unintentionally, the chimps who were baited with withheld bananas suffered the emotional strain of a food crisis caused by a natural disaster. Power says the frustrating feeding methods at Gombe induced a "stressful emotive atmosphere of a rare, acute food crisis such as might be brought about through either overpopulation or prolonged natural disaster."

The chimps plausibly perceived a crisis caused by overpopulation due to increasing numbers of researchers and human habitats encircling the park. Baiting the chimps simulated a food shortage, and added to the stress of surreal overcrowding may have led to despotism, the aggressive defense of resources. Historically, people learned little from the countless times they overpopulated their land. The food shortages that followed overpopulation produced hierarchies wherein despots confronted the issues with war or genocide or whatever was expedient. After all, it seems apparent that humans and chimpanzees differ only in degree, except the chimps didn't cause their nonadaptive dilemma.

Despotism is not sustained by the positive emotions adapted for egalitarian societies. Nevertheless, love, compassion and mercy are manifested in despotic systems even if they are weakened in emotional storms. Despotism is extreme, but forestalling starvation requires extreme measures and takes precedence over other survival means. A hypothetical example is if Evered and his cousin King Edward were famished, their motivation for pleasure would be disenchanted. To subsist, and if possible Evered would prefer bananas and King Edward champagne and lobster. However, if the two primates were in the last stages of starvation, they would eat each other.

Attempting to survive the fancied famine, the disillusioned Gombe chimps closed their borders and hurt and killed the chimps they once hugged and kissed. It would be easier for charismatic males and females to lead trusting apes away from the imagined calamity. They would fission into viable microcosms and migrate to supportive environments where they would grow into new communities. The charismatic David Greybeard, for example, might lead away uptight William and jittery Goliath along with some females, children and older chimps. Together they would find a promising land and "Be fruitful, and multiply, and replenish the earth." (Genesis 9:1)

Charismatic David Greybeard and follower

Seeking a promised land

The fruitful microcosims would soon flourish like seeds in a fertile soil. In the bipedal world Moses and his dependants are analogous to David and his followers. The alpha Hebrew emigrated from Egypt with the children of Israel toward the

Promised Land. The Israeli children were slaves of a pharaoh as authoritative as an alpha ape. Slavery isn't a chimpanzee institution, but one that presumably arose after the Agricultural Revolution when people were cursed with work. Therefore he was sent "forth from the garden of Eden, to till the ground from whence he was taken." (Genesis 3:23)

Charismatic Moses with the Children of Israel

Seeking the Promised Land

Prior to leaving for a promising land the children of Gombe needed to break up, which was difficult. Chimps are "highly attracted to the group as such, less so to specific others," Power says. Strong bonds like flexible cables hold individuals to the mobile group of self-reliant chimps. Conversely, weak bonds loosely link individuals in a society adapted for group affinity. When a friend or relative in such a society dies it's accepted as a natural conclusion, like the poet says without "funeral gloom with corpse-gazing, black raiment and graveyard grimness."

On the other hand, there's corpse-gazing and gloom in a corrupted egalitarian society wherein individuals are strongly bound to each other. Affiliated societies are supported by feelings of security that evolved with egalitarianism. Conversely, aggression and alarm that corrupted the Gombe social order produced feelings of insecurity. Responding to uncertainty, the chimps forged individual bonds of reassurance. The anomalous bonds made death a traumatic event among the anxious chimps. The death of a mother, for instance, caused her adolescent son such grief he couldn't get over it and "in a state of depression fell sick and died himself," Goodall says.

Grieving people can die from losing a loved one and consequently feel ostracized from society. Loneliness may mimic punishment for nonconformance to societal norms. Grief and loneliness have a confusing effect similar to the confusion of baited chimps. People sometimes commit suicide because they can't stand lonely and loveless lives. Shakespeare holds suicide likely results from an intolerable disappointment in intense love, romantic or familial. Juliet believed Romeo was dead and stabbed herself with a 'happy dagger.' Hamlet soliloquized to be or not to be because he was depressed about his family's intrigues.

Hamlet's late father was king of Denmark, and upon his death his widowed queen married Claudius, the dead king's brother. Claudius coveted the throne and was ambitious and enterprising like Mike, the Gombe ape aspiring to alpha status. Claudius murdered his brother by pouring poison in his ear while he snoozed in his orchard. Mike, unlike Claudius, didn't have poison to pour in an ear, and maybe not the cerebral skills to use it, but he was darn good with kerosene cans. Mike's cousin Hamlet avenged his father's murder with a scheme that killed his evil uncle.

Hamlet's solution to his father's murder was the same as the apes who murdered other apes at the Holland zoo and Florida attraction. Tragically, in his act of resolution, Hamlet died with the conclusive words, "The rest is silence." Much togetherness in a restricted society is fettered by adamant and maladaptive ties as firmly as Hamlet was bonded to his father, even in death. Clearly, lots of gloom and sorrow came when strong bonds between individuals replaced individual-group bonds in chimpanzee and human egalitarian societies. Individual-group bonds were eventually ruptured from frustration and aggression during the Gombe delusory food crisis.

Aggressive behavior is "functional when it brings about significant social change, when change is necessary for survival," Power says. "Violence, rejection and refusal of access to resources by others, formerly their supportive companions, might act as a catalyst." Aggressive behavior within a community in crisis serves to break strong emotive bonds. Therefore, aggression is adaptive to clear the way for options like migrating, forming new bands or changing the social structure to one best suited for bad times. A desperate solution in the worst of times is to fight or kill for every last fig on every dying tree.

During dire times breaking bonds and heading for a promised land with a charismatic primate is adaptive. Still, killing the dispossessed is not optimal strategy as happened at Gombe. Power says the extreme aggressive behavior of the Gombe chimps is far beyond the aggression required for them "to break away from their tense group. But as we know, human-surrounded chimpanzees no longer have this option." Imagination locked their minds in a vice for which the chimps had no adaptive response. "The mind is its own place and can make itself a Heaven of Hell, a Hell of Heaven." (Milton)

Struggling in hellish frustrations, the two chimpanzee communities began seeing each other with suspicious eyes. Their relationships grew strained and by 1972 the males met only occasionally. "Some peaceful interactions were seen, but in general the males increasingly avoided one another," Goodall says. The southerners stopped visiting camp except for Goliath and Madam Bee with her daughters who dropped by a couple of times. The tensions between the northern and southern chimps increased like they did in America from the American Revolution to the Civil War. Tariffs and other insidious economic interests divided the nation between the two wars.

CHAPTER 19

America in 1776 proclaimed itself an indivisible nation, but within a few years mounting resentment was alienating the northern and southern states. Similarly, mounting resentment at Gombe alienated the northern and southern chimps, and in 1973 they were recognized as two separate and hostile communities. The few unruffled interactions between Kasakela and Kahama males at that time were between northerners Mike and Hugo and the southerner, Goliath. Goodall says the three apes "had associated closely for many years. Indeed, Goliath and Mike had become very friendly from 1967 until the community division, and Goliath's decision to move south was puzzling."

A bipedal parallel was General Robert E. Lee of the Confederate States and General Ulysses S. Grant of the United States. The two alpha primates were graduates of West Point and had served together in the 1848 Mexican War. Grant was from Ohio and Lee from Virginia, so it's not puzzling each man affiliated with the province of his birth. Grant led the Union to victory, and in 1872 was elected President of the United States. The political system that elected him, along with other systems, arose in the last 10,000 years during the alteration of the human adaptive society.

Grant was a clever strategist in war, but as president he was like lowly ranked Jomeo and historians gave his presidency a poor grade. Grant and Jomeo had their strong and weak points and apparently different strokes for different folks also applies to chimps. Jomeo, in social skills had a lesser stroke, but in the Primal War his strokes were as effective as Grant repelling Pickett's charge at Gettysburg. General Grant, like General Lee, witnessed the deaths of thousands of men in the

Civil War, an economic conflict rationalized to attain a humane objective.

Jomeo

Superior Gombe warrior

Socially deficient

Grant

Superior American General

Politically deficient

The soldiers who fought and died in the war were born with the essential quality of love and its mellow embodiments. So were their chimpanzee kinsmen who fought and died in the Primal War. Indeed, in opposition to their positive emotions, the soldiers of the Union and Confederate armies hurled cannon balls at one another and cut each other with the sharp edges of ghastly bayonets. Indeed, in opposition to their positive emotions, the northern and southern forces of Gombe warriors hurled rocks at one another and cut each other with the sharp edges of ghastly teeth.

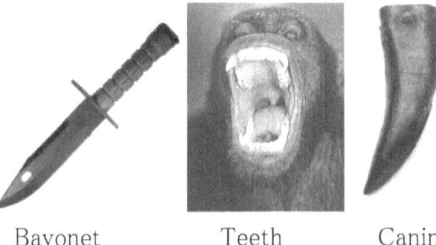

Bayonet   Teeth   Canine Tooth

Two Southern fatalities

Gettysburg soldier

Gombe warrior

The Primal War commenced without warning in 1973 with the sudden assault on the male chimpanzee Godi in the green forests of Gombe, Tanzania. Similarly, the Civil War began without warning in 1861 with the sudden assault on Fort Sumter in the sapphire waters of the harbor of Charleston, South Carolina. America was a house divided against itself that stood as one nation after a horrible war. On the other hand, the Gombe house divided against itself perished from this earth with the death of the southern male Sniff in the final conflict in 1977.

The tragic war ended with the annihilation of the southerners by the northerners who witnessed a Precivilization Gone with the Wind. The fall of the house of southern Gombe began in the fated year of 1973 when a cadre of Kasakela combatants suddenly, and without warning surrounded Godi, an unwary male feeding in a tree. The frightened chimp leapt from his limb and fled for his life, but was quickly grabbed and slammed on the ground. The frenzied apes pinned him down while one sat on his head and held his legs together.

The enraged assailants pounded him unmercifully on his back and shoulder blades with their hands and fists. They bit him and threw rocks at the helpless southerner who didn't have a chance to escape the horrid mauling. Goodall says throughout the attack all the chimps "had been screaming loudly." Their screaming and hooting split through the Gombe forests like the shrieking of Yankee soldiers cleaved the air as Sherman assaulted the southern citadel of Atlanta. Goodall says Godi had "a great gash extended from his lower lip down the left side of his chin, and his upper lip was swollen.

Kasakela assailants going after Godi

Declaration of War

"He was bleeding from his nose and from cuts in the side of his mouth. There were puncture marks on his right leg and between his ribs on the right side, and he had a few small wounds on his left forearm." Godi was reported missing in action and wasn't seen again and presumed dead. Some other MIAs were found or detected by the stench as often happened when a Confederate or Union soldier crawled off and died. The soldiers that were never found were some of the 600,000 Johnny Rebs and Yanks who didn't come marching home again.

"One boy on whose face a blond fuzz had just begun to sprout, was dumped on the front porch by a mounted soldier bound for Fayetteville," Mitchell says. "The girls thought he must be one of the little cadets who had been called out of military school when Sherman approached Milledgeville but they never knew, for he died without regaining consciousness and a search of his pockets yielded no information. Somewhere to the south some woman was watching the roads, wondering where he was and when he was coming home.

"They buried the cadet in the family burying ground next to the three little O'Hara boys. And Melanie cried sharply as Pork filled in the grave, wondering in her heart if strangers were doing this same thing to the tall body of Ashley." The reasons for such sad stories as the little cadet are plausibly found in the human and chimpanzee genotype. Goodall says chimpanzees stand at the "threshold of human achievement in destruction, cruelty, and planned intergroup conflict." Coded by their nearly identical DNA, the two primates share a spectrum of positive and negative emotions from love to hate.

Goodall says chimpanzees show compassion when they help and sooth "those in distress. They are almost certainly capable of feelings akin to sympathy." The negative emotions emerge when they kill other chimps who might be their relatives or former friends. Goodall says before the Gombe communities split, the chimps had "enjoyed close and friendly relations with their aggressors." Still, when they waged war against their friends and relatives, or harmed unfamiliar chimps, they suppressed their positive emotions. Likewise, Union and Confederate soldiers that fought against their friends and relatives from neighboring states subdued their positive emotions.

Friends and relatives within the frustrated Kahama and Kalesekle communities fought without killing each other, but among 'civilized' people conflicts reach new levels. Husbands and wives who vowed to abide with each other until parted by death have hastened the parting by murder. An infamous case of fratricide involved a vegetable farmer and his brother who herded sheep. The brothers gave their mentor a fruit basket and a meat portion respectively. The mentor scorned the fruit in preference to the meat. Frustrated and in a jealous rage "Cain rose up against Abel his brother, and slew him." (Genesis 4:8)

Cain wasn't stoned to death for his capital crime and later married and begat children. Provided the genealogy of Cain's parents is factual, executing Cain might have discouraged them from begetting more children. A barren genealogy of the first couple would have prevented Homo sapiens from populating the earth and spared countless species from extinction, including the Kahama chimps. In the bipedal world those guiltless chimps could have been fraternity brothers and sorority sisters to the neighbors who killed them. All they needed were bigger brains, less hair and a lot less testosterone.

The triumphant Kasakela apes and vanquished Kahama apes had played together as children, romped over the forest trails, fed in the same trees and met gleefully at carnivals. When they were adolescents they shared copulatory pleasures with the same eager partners. Indeed, perhaps Madam Bee who was brutalized and killed by Kasakela warriors was the southern belle whose hand was gently kissed by a genteel northerner at the jovial 1960 Festival of Figs. Gombe chimpanzees and Americans that fought in their respective wars had different objectives and rules of warfare.

Madam Bee

Kissed and later killed by her Kisser??

When Civil War soldiers surrendered in battles, both sides took captives, treated injuries and sometimes exchanged prisoners. The objective of General William Tecumseh Sherman in 1864 as he marched through Georgia from Atlanta to Savannah was to break the economic back of the South. Neither side intended to annihilate the populations of its adversary. Conversely, during each chimpanzee attack "the observers, all thoroughly experienced in chimpanzee behavior, believed that the aggressors were trying to kill their victims," Goodall says. Since all the southerners were killed, it appears the observers hit the nail on the head.

Goodall asked the observers why they thought the aggressors were trying to kill their victims. The observers said the attackers showed patterns seen when large prey was killed that were not seen during fighting within the community. The Kasakela objective had the same objective as wars that are planned to exterminate a population, or eradicate an ethnic group by genocide within a nation. "And we took all his cities at that time, and utterly destroyed the men, and the women, and the little ones, of every city, we left none to remain." (Deuteronomy 2:2)

Each hostile chimp was as savage in wasting a relative or former friend as Samuel was when he "hewed Agag in pieces before the Lord in Gilgal." (1 Samuel 15:33) Hewing Agag to pieces or brutalizing Madam Bee were aggressive acts perpetrated on victims of the same species. Frustrated chimps in cages have escaped and attacked people and almost killed them. Disturbed chimps in Africa whose adaptive method of fissioning is restricted, have killed unwary people. The negative emotion of hatred and its component frustration cause chimps to kill chimps and people to kill other people like poor Agag.

The bonds of group affinity inherent in humans and chimpanzees are supported by the positive emotion of love and its components. When those bonds are broken, it is done by negative emotions such as hatred, jealously, anger and aggression. Intense negative emotions initiate pseudospeciation in extreme situations like warfare when sentient chimpanzees or humans kill relatives or erstwhile friends. Pseudospeciation, Goodall says, signifies that "the members of one group may not only see themselves as different from members of another, but also behave in different ways to group and non group individuals."

A psychologist coined pseudospeciation, or false species, which means to perceive a member of the same species, or conspecific as another species. Pseudospeciation is a rationalization denoting people can imagine vast differences in others when the differences are none or minor. Poor Agag had 99.9% the same DNA as Samuel who knew Agag was a fellow human. However, in their nonadaptive society Samuel viewed Agag with a prejudicial eye. Samuel considered Agag a threat and 'pseudospeciated' him as he hacked away at the unlucky man. When pseudospeciation inhibits cooperation between groups or individuals it fosters conflict and mistrust.

Samuel pseudospeciating and hewing Agag to pieces

A nonadaptive situation

The milder function of pseudospeciation is to transmit "individually acquired behavior from generation to generation within a particular group, leading to the customs and traditions of that group. The process is analogous to a species forming through genetic inheritance," Goodall says. Pseudospeciation is perhaps Balwinized in the genotype, and if so, it is an adaption of a social system structured on group affinity. Pseudospeciation responds to the situation and can be mild or intense. Severe pseudospeciation might be a corruption of an initial adaptation and would hardly be utilized by chimpanzees in a world free of human predators.

Pseudospeciation and rupturing bonds serve in unusual situations such as a natural disaster. In a disaster, chimpanzee communities would break up into microcosms and migrate with the guidance of charismatic apes. They would be loath to employ extreme pseudospeciation and kill their neighbors, and instead fission and avoid genocide. Killing conspecifics is maladaptive since their birth and death rate were optimized over many millennia to remain relatively stable. What's more, dead chimps don't pass on their genes. Human genocides are not adaptive in a system of indirect competition where resources are occasionally scarce, but never completely exhausted.

Humans and chimpanzees kill conspecifics when their adaptive society is inadvertently or unwittingly altered. People dehumanize other people to oppose them, and more drastically to kill them. So much for love thy neighbor. Pseudospeciation lets people devalue different nationalities, ethnicities, races, religions or contradicting opinions. Then too, pseudospeciation might be a chimpanzee adapation coded in humans by the same genes, but magnified and distorted in the tumorous brain. During warfare those 'others' or 'them' are dehumanized and any atrocity is justified. Likewise, Goodall says in chimpanzee warfare opposing chimps "are 'dechimpized' and treated as though they were prey animals."

Dechimpizing or dehumanizing suggests the milder role of pseudospeciation to transmit cultural behavior is adaptive and can be intensified in extreme situations. Still, extreme pseudospeciation might be a corruption of its original adaptation. When a pet chimp is frustrated and attacks a familiar person it's an extreme use of pseudospeciation. Primary adaptations recruited for a secondary function is similar to using cranial vessels adapted to counteract gravitational restraints to cool enlarging brains. Whether chimpanzees ever employed the intensity of secondary pseudospeciation during a natural catastrophe before their society was corrupted is unknown.

A modification of an adaptation like pseudospeciation is beneficial when it aids survival and reproduction. However, conceptually it's maladaptive, sub–adaptive or nonadaptive when it is detrimental. Indeed, severe pseudospeciation might be a mechanism that strains humans and chimpanzees against the grain of their genetic directives. In other words, it may not have been selected for murder or killing in warfare although it does the job. Two strains against the grain are the big brain and consequences of direct competition that could lead to the obliteration of humans. A frequent prophecy for that calamity is a nuclear or biological war.

Certainly, the obliteration of Homo sapiens would argue that the big brain and direct competition are maladaptive. Provided people are obliterated, most of them have a survival mechanism in a religion that entails certain conditions. One is to practice the Golden Rule and dispense good deeds to others in hope others will return the good deeds. Undisturbed chimpanzees practice an unspoken version of the Golden Rule to live and let live. However, chimps and people suspend the rule during warfare instead of dispensing good deeds, they pseudospeciate those 'others'.

Pseudospeciation has limited uses by chimpanzees, but humans have more systems and more opportunities to debase their fellows. Evolutionists have been pseudospeciated by their detractors since Darwin published the *Origin of the Species* in 1859. Pseudospeciation plays subtle roles such as passive racial and religious discrimination. Since strangers are different they may be feared, a condition termed xenophobia, and be pseudospeciated. Three Gombe chimps were partially paralyzed during a polio epidemic and were subjected to xenophobia. When the other chimps saw the cripples they were extremely fearful, and as their fear decreased they became aggressive.

Many of the chimps "displayed toward and even hit the victims," Goodall says. Things that are different or unknown have the potential to be harmful, and being apprehensive is instinctive. Clearly, the snap of a twig or not respecting an unfamiliar animal can quickly spell doom. Equally bad is the click of a rifle that has spelled doom for many a chimp. A puzzling situation with a group of baited chimps is attacks on stranger females ending in the deaths and cannibalization of their infants. The attacks were possibly instigated by xenophobia, the fear of strangers.

Goodall says in twelve "gang attacks on mother-infant pairs, twice infants were seized, flailed, and eaten; twice they were seized, flailed and hurled away; twice they left their mother and climbed a tree, where they were ignored; once an infant left the mother and was attacked, then left; and on the other five occasions they were left clinging to their mother during the fights." Curiously, all weren't killed, or only partially eaten and the mothers weren't recruited for mates. Goodall surmised the attacks were directed at the mothers because they were strangers and chimps "have an inherent aversion to strangers."

This "behavior seems to me bizarre," Power says. "That is, extraordinary, and involving striking incongruities, maladaptive and pathological. Social tension may cause chimpanzee behavior to spark over temporarily to infanticide and cannibalism from the standard opportunist pattern of predation of chimpanzees." Obviously, such violence is contrary to the peaceful 1960 Fig Festival that was filled with merriment, hugs and kisses. Raising the number of infants to adulthood that a foraging range can support is crucial for evolutionary success. The Gombe and provisioned chimps have amble ranges and killing to control populations is unwarranted.

Curiously, angry male chimps were seen beating up mothers and their children and later were affectionate and protective with the children. They acted like abusive men who harm their family and later feel remorse and plead for forgiveness. Those poor mothers in the bipedal world would be in a battered women's shelter. Certainly, something disturbed the chimps after the exuberant Fig Festival like something disturbed people during the Agricultural Revolution. People and chimps evolved late on the geologic clock and have virtually the same DNA. They are made of carbon, hydrogen, oxygen, nitrogen and other elements present in extraterrestrial objects.

The elements were components of a mass of gases that drifted through space and contracted to form the solar system. The condensing gases grew hot and when the earth cooled it had the potential to sustain life. Whatever the essence of living matter on earth is, it originated in the Precambrian Period with organisms that responded by instinct. When they and animals that are aware of existence die, their life processes cease and they decompose into their rude elements. "Dust thou art, and unto dust thou shalt return." (Genesis 4:19) Certainly, this is the end.

CHAPTER 20

There is continuity and discontinuity during the evolution of life from the Precambrian Period to the end of the Holocene Epoch. Organisms evolve from their precursors randomly into bushes and become new species or meet extinction. The written chronology of human evolution begins with inanimate particles that fused into living substances, whatever defines life. Transitional words and phrases at the beginning, end, and within each paragraph aim to keep the train of thought flowing smoothly. Conversely, stream of consciousness writing expresses spontaneous thoughts. James Joyce uses the style of stream of consciousness in his epic 1922 novel *Ulysses.*

Off the cuff, in every population there are a few people with an unusual capacity for kindness, generosity and cooperativeness. Mayr says the families of these good people always insist they have "been that way from infancy on." Their altruism is as positive in one direction as the conscience of sociopaths is negative in the opposite direction. Most people fall between the extremes in the middle analogous to the average chimpanzees between charismatic and nervous individuals. Credibly, an unusual capacity for kindness is adaptive in specific individuals to foster group affinity among bands of human and chimpanzee foragers.

Some friendly individuals are best at promoting congeniality in meetings and carnivals. On the other hand, tough individuals could be adapted to enforce ostracism on deviants. Human adaptions that are influenced by like genes in the chimpanzee genotype might be magnified by the big brain or in the person's childhood. Some people are cruel, for example, which could be the characteristic of toughness intensified by a harsh childhood and overblown by the big brain. In similar manner, altruism and charisma are conceivably enhanced by a loving family, or subdued by a callous upbringing.

The altruistic David Greybeard "was very generous and tolerant during banana feeding," Goodall says, sharing his fruits with females and youngsters. He was fearless, and in 1964 led the Gombe partisans in overthrowing Mike, kerosene can tumbler. David's bipedal parallel is a caring person who sacrifices for others. The apostle Paul says without love I am like "sounding brass, or a tinkling cymbal. And though I have the gift of prophecy, and understand all mysteries, and all knowledge; and though I receive all faith, so that I could move mountains, and have not love, I am nothing." (Corinthians 1:13)

Anthropologists are cautious when drawing behavioral parallels between human and chimpanzees. Still, their uncanny behavioral similarities imply they were seeded together in a nutshell of evolution. Gombe despot Mike and the charismatic Nazi dictator Adolph Hitler manifested a sense of insecurity. Hitler's insecurity possibly developed in his childhood from being frequently beaten by his frustrated father. Power says "it is those (animals or humans) who are insecure and most in need of status who are most likely to be the first to behave aggressively, and so gain hierarchal status." Hitler aggressively used his resources in a relentless rise to power.

Oppression

Adolph Hitler

Alpha Despot

Constantly plotting lest
he be disposed

Gombe Despot

Constantly plotting lest
he be disposed

Mike in his relentless rise to power was aggressive with the
clamorous kerosene cans. Eventually, despots Mike and
Hitler were deposed, despite their plotting. Humphrey
toppled Mike and Hitler committed suicide after being

defeated in war. Mike and Hitler lived in nonadaptive societies wherein both struggled and schemed to achieve and protect their tenuous position. Hitler became an absolute dictator, but was always uneasy because of attempts to assassinate him. Hitler's Machiavellian stratagems to solidify power and protect his position were in vain. De Waal says "Whole passages of Machiavelli seem to be directly applicable to chimpanzee behavior."

One of Mike's stratagems was spending "more time than is usual in a high-ranking male to the grooming of subordinates," Goodall says. Perhaps in a foraging society the insecure Mike would have been an alerting chimp in the structure of attention. Hitler, in a hunter-gatherer tribe with his artistic aptitude might have painted the face of a medicine man. Circumstances destroyed any potential the two primates had to live and be productive in their natural social organization. Credibly, in his nonadaptive situation, the abuse Alois Hitler inflicted on his son Adolph incited a drive for power in the young Austrian.

Conversely, Flint, a pampered son at Gombe was born with a silver banana in his mouth. His mother, Flo, gave birth to the coddled ape in 1964 during the baiting project. Power says the tense social atmosphere induced by the project caused changes such as prolonged dependency and deficient social maturation. A notable change was in weaning that previously was started by nursing infants. Weaning became a toilsome task the nursing mothers had to enforce. Goodall says some "weaning is accompanied by frequent aggressive incidents," exemplified by Flint who was hard to wean.

The spoiled Flint rode on his mother's back and shared "her nest at night, when he was eight years old," Goodall says. Apes mature before people and when Flint died at eight and a half years old, his behavior was arrested. Another arrested adaptation was males wouldn't emigrate and youthful females delayed emigration. Their reluctance to leave was plausibly based on insecurity and not a desire to stay. Power says the delay to transfer is "beyond the age of independence characteristic of wild chimpanzees." Flint would hardly want to go anywhere because he was so dependent on his indulgent mother.

Flint piggy-backing on Flo

Flint was the son who died from depression after his mother's death. Ironically it might seem, his older brother Figan became alpha ape. Figan was born before the baiting project and was an adolescent when the project began. During his adolescence he endured the alternating hostile and protective behavior of big males. The subsequent aggression that baiting provoked was not so much a struggle for rank, Power says, as it was redirected aggression, scapegoating "and such well-established frustration responses,-in that many of those attacked are females, children and elderly chimpanzees, noncontestants in the power struggle."

"Chimpanzee infants," Goodall says, are exposed to aggressive acts "from a very early age. Sometimes this is at first hand, for even a mother carrying a tiny baby may be attacked, and although she typically crouches over it at such times, the baby may be hurt." Power says a human child who feels confidant enough to leave its parents is likely to leave "earlier if its experience has been with a familiar, nonthreatening world, than if it perceives the world as threatening." In modern society many adults, at least in age, live with their parents for various reasons.

Many adult 'adolescents' can't cope with competitive society because they haven't been effectively 'weaned.' Numerous homeless people didn't have the training to function in a nonadaptive economic system. Sadly, lots of homeless people are mentally ill and 'pseudospeciated' by callous people. Flint would be homeless in the bipedal world without his mother, but if his brother Figan were a man he would have the potential to be successful. During Figan's quest for power, like Mike and Hitler, he quickly took advantage of weaknesses. The hardhearted ape exploited the partial paralysis of his older brother Faben, stricken by polio in 1966.

When Figan was an adolescent, Goodall says, his charismatic mother helped him "in victories over other young males." Figan was an excitable challenger and the chimp who clutched his own genitals which "seemed to indicate a lack of self-confidence." Provided lacking self-confidence and insecurity are overcome by ruthless ambition, suggests Adolph Hitler also clutched at his own groin. Figan was highly motivated, and after he defeated Faben, his partially paralyzed brother, the forgiving Faben supported him in becoming alpha ape. The coalition of Figan and Faben defeated Humphrey and sent the challenging Casanova Evered back to his lustful pursuits.

When Faben died, Figan became friendly with his former adversary Humphrey to solicit the support of the big ape.

When Humphrey died, Figan wasted no time and became close to the romancing Evered and the affable ape, Jomeo. Despite Figan's power, he had fewer sensual delights than Evered and less downtime than Jomeo. When Figan rose to alpha position in 1972 it was two years before first blood was spilled in the war between the communities. War rumblings and tensions were permeating the air and once friendly neighbors eyed each other with distrust.

The apes were as uptight as many people living in the tense atmosphere of materialism. Sometimes their subdued frustrations are redirected in aggressive incidents such as road rage, spouse abuse and kicked dogs. Subdued frustration at Gombe seethed in silent squads of tense males patrolling their community borders. Any strangers or relatives or erstwhile friends daring to approach were assailed violently. Power postulates the patrols were distortions of chimps looking for a carnival. Moreover, when jocular chimps seeking carnival are attacked it may cause a bellicose domino effect gusting through the Gombe communities. Indeed, an ill wind blows no good.

When the Kasakela ill wind blew the Kahama southern chimps to oblivion the victors "began to sleep as well as feed in the area which, for the previous five years had been the heart of the Kahama core area. But this situation did not last for long," Goodall says. What goes around comes around and the ill wind shifted. The vanquished Kahama community had buffered the Kasakela chimps from an outlying community. Soon the outsiders encroached and the captured area "began to shrink." Further, another baited study distant from Gombe showed "a similar aggressive relationship between males of different communities."

Meanwhile, one woe followed another woe and more human interference agitated the already agitated chimps. During a civil conflict in 1975 there was an "unfortunate kidnapping event" of several student researchers that were later released, Goodall says. Surely, for the chimps it was like H.G. Wells' *The War of the Worlds*. War, aggression, jealously and hate fascinate people when viewed through detached events. Sleazy television shows with women attacking their rivals elicit audience excitement. Children love horror stories like Hansel and Gretel who baked an old woman in an oven, a fairy tale as dreadful as the Gombe killings.

The Gombe killings and the occasional violence of pet chimps give the impression chimps are inherently aggressive and male dominance struggles support the notion. However, Goodall says male chimps are mostly "relaxed in one another's company, and spend a good deal of time grooming, feeding and traveling together." Chimps are fundamentally peaceful apes, but the perversion of their social principles "spread in ripple fashion to distort all aspects of the adapted social order," Power says.

Likewise, when people began planting seeds and domesticating animals all aspects of their adaptive social order were distorted, commingling tranquility with distress.

Tranquility is enhanced by love and the properties that sustain the egalitarian society. One time when a tranquil chimp was frightened, she restored her tranquility by "physical contact with a companion by touching, embracing, mounting, or kissing," Goodall says. Two female chimps embraced and shrieked for joy when they suddenly saw a big pile of bananas. Faben was so excited when banana boxes were opened he turned his rump toward a companion. When his companion reached to comfort him, Figan made thrusting movements and bounced "his genitals up and down against his comrade's palm."

Certainly, a handful of prodigious testicles are sexually suggestive, but Figan thrusting and bouncing his gonads only served to relieve tension. The need for comforting physical contact by chimpanzees and humans in emotional moments indicates common genes direct the action. During their adolescence Figan and Pooch plausibly practiced the comforting intimacy of sexual intercourse. They went off on what was "for each of them, almost certainly the first consortship," Goodall says. The time the young lovers surrendered their virginity is comparable to the time high school boys and girls surrender their virginity, at the first opportune moment.

Pooch, like Figan, was an adolescent in the tense period of baiting shortly before two Gombe communities were recognized. Mesmerized by the gilded treasure, the baited apes were frustrated and nervously anticipating the next box to pop open. Sometimes Figan and Pooch were scapegoats for redirected aggression emanating from frustration, the usual reason for people scapegoating. Figan and Pooch and other chimps avoided unpleasantness by romancing alone in a private retreat. Similarly, heterosexual, bisexual, and homosexual people rendezvous privately when necessary to avoid unpleasant situations. The chimpanzee secluded interludes bolster the hypothesis that their consortships result from distorted social principles.

Figan and Pooch spent most of their formative years in post 1965 Gombe. During their teens the fundamentals of their egalitarian society slowly altered. Indirect competition, fission and fusion in flexible subgroups of confident-reliant apes gave way to direct competition and restricted movement. Hoards of anxious Gombe primates abandoned their principles and hastened to fight over golden bananas. Contributing to their confusion were additional researchers and indigenous people surrounding the reserve. One bewildering instance was young Figan commandeered Mike's discarded kerosene cans to practice with them alone "in the bushes," Goodall says. When the cans were removed Figan was baffled.

Equally baffling to their large-brain relatives was their social structure was distorted 10,000 years ago. Modern economic systems that developed after indirect competition switched to direct competition are profoundly intricate. Karl Marx used mathematical formulas in his voluminous publications to explain the complicated functions of capitalistic production, consumption and services. Credibly, the chimpanzee genotype is adapted for consumption solely, unless grooming is viewed as a service. However, chimps can be coaxed to barter in the artificial setting of lab experiments. Deductively, in examining the similitude of the human and chimpanzee genomes, people are not adapted for production or services.

Thus, if so, capitalism or any moneyed system is nonadaptive and as artificial as lab chimps trading tokens for treats. Sustaining egalitarianism, even with laws derived from documents such as the Magna Carta and the U.S. Constitution, indicates directly competitive human economic systems are as unlikely to change as leopards changing their spots. John Stuart Mill's Principles of Political Economy evaluated capitalistic production, consumption and services. Mill's work was published in 1848 coincidentally with Karl Marx's The Communist Manifesto. The two economic theorists write that human societies were simplistic before the emergence of complicated economic systems.

A simple society would spare many affluent middle-aged men an untimely end since they have the highest suicide rate in capitalistic societies. Another sad consequence that arose with directly competitive systems was slavery, a thorn in the side of human compassion. However, the economic motive for slavery is so strong that indifference to the inequality of servitude interferes with altruism. Christendom's Judas

Iscariot exemplified the power of money. Judas asked the priests, "What will ye give me, and I will deliver him unto you?" (Matthew 26:15) Judas identified his redeemer with the Kiss of Death for thirty pieces of silver.

Kiss of Death

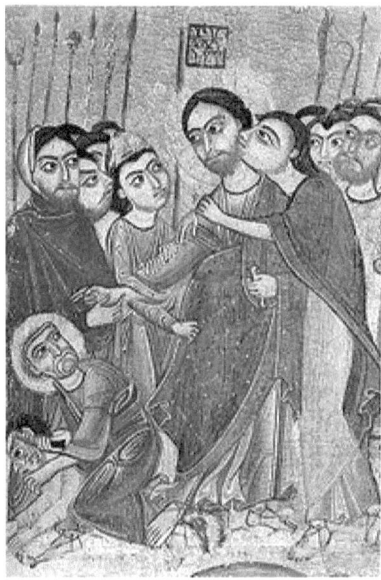

Betrayed for 30 pieces of silver

Thirty pieces of silver bought a lot in olden days, such as indentured servants or slaves. Since historic records are imprecise, researchers deduce that subservient and mutually supportive institutions such as slavery and marriage respectively, arose during egalitarian social changes. The first slaves were captured in tribal wars over crops or land, and as the institution grew, they were bartered or sold. Compassion necessitates pseudospeciating slaves to see them as those 'others,' but maintaining slavery conflicts with the helping behavior of altruism. Goodall says to maximize reproductive gain "helping behaviors should be directed exclusively toward kin, the closer the better."

Thus, it is easier to pseudospeciate slaves or any group the more distant they are in kinship, cultural and physical characteristics such as language and color. African slaves, for example, were typically bought and sold to people of different cultures and colors. Rationalizing slavery, pseudospeciating slaves assigned them to another species, but sometimes it didn't work. Goodall says "helping behaviors will on occasion be extended to familiar individuals even when they are not very close kin." Therefore, when people and their slaves associate they may form bonds emanating from the genetic propensity of group affiliation, shades of the Stockholm syndrome.

Slaves of the American south weren't permitted to be educated, which discouraged bonding and separated them as those 'others'. Provided familiarity breeds contempt, familiarity is not good for controlling slaves. Some slaves were given, or allowed to buy their freedom because they became friendly with their owners. Bonding through familiarity degraded the purpose of a slave in a ranked economic system. A supervisor, especially a pragmatic one, between master and slave best served the institution. The pragmatic measure of cruelty instilled fear into slaves subdued in an institution thought to be one result of the Agricultural Revolution.

Two slaves

CHAPTER 21

Cruelty is conceivably an adaptation in a crisis to break bonds of group affiliation or it's a response to situations humans and chimpanzees have no adaptive response. Still, in the worst of natural disasters as simulated at Gombe, cruelty would help destroy a society by killing some members, and leaving the survivors the remaining resources. Whatever cruelty's purpose, if any, it debases loyalty and ruptures bonds that were selected for the greater good. Loyalty, an egalitarian property between slave and owner would plausibly promote more efficient productivity. Mill says, "Hired labor is generally so much more efficient than slave labor."

Therefore, hiring workers can be cheaper than housing and feeding slaves whose work is less productive since the profit motivation is absent. The Industrial Revolution with its invisible hand made it economically advantageous for some nations to liberate slaves, a back door entrance for the Golden Rule. Treating slaves humanely is like drawing more flies with honey than vinegar. Perhaps slavery, that Mill called a curse and "nefarious enterprise" would have lasted longer if the slaves were treated better. Treating adults badly doesn't necessarily make them cruel, but cruelty might result from childhood disturbances.

"Even when no obvious behavioral abnormalities are apparent in the adult, evidence is accumulating which suggests that certain vital functions may be seriously and permanently affected by disturbances in the caretaker–child relationship during infancy," Goodall says. Disturbances are common in adults raised with little or no affection, which is a positive egalitarian property that was displayed by the hugging and kissing chimpanzees at the 1960 Fig Festival. The larger human brain magnifies the positive and negative emotions promoting great humanitarian deeds or inhumane acts. Compassionate people will always suffer from the misery that never ceases in a maladaptive society.

Goodall tells of love, compassion, generosity, support, tolerance and friendliness exhibited by chimpanzees. Once Madam Bee was ill and couldn't climb a tree to feed with her daughters Little Bee and Honey Bee. Little Bee climbed down and placed a fruit "on the ground beside her mother. She then sat nearby and the two females ate together." When a captive male was without food for several days he appealed to an elderly female. "Eventually she responded by gathering a pile of food and taking it over to him." The Good Samaritan chimp fed him for five successive evenings.

The Good Samaritan

Blessed are the merciful

Another happy ending was the time a young female named Cindy fell into a moat and sank. An unrelated female adolescent "stepped into the water and grabbed one of Cindy's arms as the infant surfaced again." Since chimps can't swim, the rescue could have cost Cindy her life. The female Lucy lived with people and if they became ill "Lucy would try to comfort them by kissing and putting her arms around them." Two captive chimps were locked outside in a cold rain. When their keeper let them in they showed appreciation by hugging him "in a frenzy of joy."

Although the big hoodlum Humphrey looked and acted like the dictator Benito Mussolini, Humphry was protective of the elder Mr. McGreggor who was crippled by polio. Chimps adopt children orphaned at their mother's death and raise them to maturity if they aren't too young to survive. Compassion is a feeling adapted to sustain the egalitarian society and is venerated in human societies. Chimpanzees and humans may be compassionate or unmerciful according to circumstances. Physically, the two primates are easy to tell apart, which is ironical since chimps are genetically closer to people than to gorillas.

Orphaned chimp

Suffer the little children

Chimpanzees and humans should resemble each other as much as chimpanzees resemble bonobos, except for the distortion of the human phenotype by a juvenile endocrine system. People are genetically 99.9% alike and the miniscule difference accounts for superficial traits. Human races are distinguished by minor physical characteristics, but mainly by color. Race is defined as an adaptation to adjust phenotypes to new or changing environments. Consensus holds dark people living in northern climates will eventually adapt lighter skins and vice-versa. Further, white African descendants in Europe had lightened in stages as they settled temporarily in transit and later moved on.

Africans that settled part way in Mediterranean climates adapted olive skins and those settling in northern Europe became white. Some populations remaining in Africa were isolated and aren't as genetically close as they are to some white people. The nearer kinship between the white and dark people is because the whites descended directly from the isolated dark population, their nearest African kin. The white population was separated a shorter time from its dark progenitor than from the African populations that didn't emigrate. Regardless of the difference among any of the populations, their DNA is 99.9% the same.

Mediterranean Sea

Mediterranean people adapted olive skin

Hypothetically, if some African emigrants took 50,000 years to turn white while the remaining African populations were isolated for 200,000 years, the isolated Africans had 150,000 years to become slightly different genetically. The negligible difference in new and co-opted old genes garnered during emigration determined the five racial colors of black, brown, red, yellow and white. Co-dominant genes code skin colors whereas dominant and recessive genes code eye colors. Every person has a unique genotype that is nearly the same as other human genotypes. The uniqueness accounts for individual differences of phenotypes and behaviors, including the degree of personal egalitarianism.

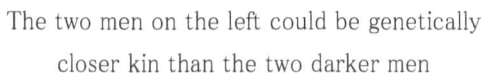

The two men on the left could be genetically closer kin than the two darker men

Humans are born with the positive emotion of love and its components that were stifled in the white European Adolph Hitler, the dark African Idi Amin and the yellow Oriental Pol Pot. These murderous despots were controlled by the negative emotions of hate, jealously, anger and vengeance. Perhaps their wrathful deeds festered from a distorted adaptation to ostracize deviants in the original human society. The genetic disposition and formative years of the tyrants justified their intense pseudospeciation of fellow humans as those 'others.' Pseudospeciation appears to be a chimpanzee characteristic inherited by humans and overblown by the lopsided big brain.

People need guidance and training during their early years to be successful in a directly competitive society. Certainly, the property of love is important in child rearing, but love alone isn't enough to prepare adults for success. Moreover, success or failure can be affected by individual differences that are determined by genes. Neglected, abused and latch key kids or other negative conditions strike against conformance to behavioral norms. Improper training and bad experiences during childhood and adolescence plus individual differences may cause people to be manipulated by capitalism's invisible hand and be led into financial difficulty.

When people work excessively to elevate and maintain a superfluous lifestyle, the bonds of group affiliation may be strained or broken. Unrestrained spending overheats the economy and keeps noses to the grindstone as corporate mills grind down unrenewable resources. Preparation to succeed in a directly competitive society opposes directives of the adaptive genotype. The best preparation requires lengthy schooling or apprenticeships that are sometimes augmented by luck. Segments of the society, like those who can't hold a job or homeless people, find life difficult for various reasons such as poor raising, unsuited dispositions or mental problems.

Some of these unfortunate people become criminals that constitute a large class responsible for millions of dead bolts and security systems. Driven by the invisible hand of capitalism, or political economy, the hopeless drug war has created its own economic system with agencies, counselors, law enforcement and prisons overcrowded with nonviolent inmates. Ironically, and sadly, the criminals, like other troubled people are born with the essential properties of love and its components that maintain egalitarianism. Sadly, but not ironically, bipedalism preceded the big brain that assures the continuance of the maladaptive society it inadvertently corrupted from the original hunter-gatherer society.

The adaptive society of hunter-gatherer peoples and their undisturbed chimpanzee cousins was a smoothly operating system. Since unmolested hunter-gatherer and chimpanzee societies no longer exist, they are evaluated by data from earlier studies. Goodall said in 1986 if chimps survive for a while in freedom it will be in a few isolated patches of forests "where opportunities for genetic exchange between different social groups will be limited or impossible; this is already the case at Gombe." Indeed, those chimps may forever be in the clutches of an imaginary crisis, unless there is a salvaging miracle of human extinction.

Adolescent chimps and hunter-gatherers in their carefree days were not born expert hunters or foragers, but learned their skills from adults. Young females of both species acquired mothering techniques by 'aunting' or helping mothers with their babies. Learning needed skills, and transcending from adolescence to adulthood, are facilitated by strong bonds of group affiliation. Conversely, in nonadaptive societies that aren't egalitarian, transitions are difficult. Goodall says after the Kahama community was annihilated females continued to stay close to their mothers and early adolescent males spent more time with adult males but with "increasing wariness of them."

The male wariness and mother-daughter tight bonding were supposedly caused by stresses of direct competition over bait plus continuing human encroachment. Adolescence in some human societies is long and difficult Power says, "and almost missing in others. As we Westerners understand the term, adolescence has come to imply generational conflict, some rebellion of the young against the adults, some exclusion of the young adults." Pushed by the invisible hand of profit motive, the stresses of Western society polarize families and many people have trouble growing up. Indeed, emotional immaturity might contribute to the rise of sexual aberrations, notably pedophilia.

The arrest of emotional maturity during the formative years could account for the adult pedophile's juvenile perception of sex and direct the sexual urge toward children. Provided pedophilia is an infantile sexual perception of an intelligent person suggests some chimps might be perverts if they had higher IQs. Certainly, the affable ape Jomeo with his scoutmaster's disposition and a following of admiring young males had the perfect opportunity for sexual deviancy. However, Jomeo could never be a pervert even with the IQ of a genius if it's a disproportionate brain that initiates perverted behavior.

Another chimpanzee behavior no longer practiced by people is the healthful and social custom of grooming. Chimps groom themselves and others to remove parasites whereas they groom socially to reinforce group bonds and restore harmony after discordant situations. Grooming is comparable to people pleasantly chitchatting or reconciling discord through discourse. Plausibly, the ballooned brain and not natural selection inadvertently produced speech and the juvenile endocrine system that limits hair growth. Since people have little hair to hide bugs and are able to converse with each other, Homo hominid grooming credibly ceased to be adaptive sometime during brain expansion.

Grooming is adaptive for physical and mental health while copulation is a reproductive adaption to promote species survival. Although grooming appears a minor function, it conforms to the definition of an adaptation that adds to the fitness of an organism. When adult chimps accept adolescents for grooming and sexual partners, it signals their transition into adulthood. Conversely, in ranked societies like post 1965 Gombe and typical Western World societies the transition into adulthood is difficult and sometimes arrested. Flint never grew up since he was an emotionally immature chimp like his many Peter Pan cousins in the bipedal world.

Flint's mother was born into an egalitarian society that was not hierarchical. Undisturbed chimpanzees had an implied structure of attention composed of restive, discerning and decisive apes with human counterparts. The restive apes pass information to the discerning apes who determine its worth and dismiss it or send it to the charismatic chimps to decide a course of action. Normally, the control element of the charismatic leaders is muted and appears "as willingly accepted," Power says, "but otherwise unenforceable, influence." Charismatic elder chimps that have 'retired' retain their admired status like respected teachers attending a class reunion.

Charismatic leaders in human foraging societies aren't distinguished by any single personality type or characteristic. Leaders aren't "arrogant, overbearing, boastful, aloof or acquisitive," Power says. The charismatic David Greybeard exemplifies the absence of negative features seen in pushy apes like Goliath and Humphrey. David in a foraging group of the immediate-return society would be the respected figure occasionally changing roles prescribed by the situation. The admired figure in the adaptive human society would be a charismatic person occasionally changing situational roles. A skillful hunter, for example, might tend a wound in the absence of a knowledgeable healer.

David Greybeard's ideal counterpart in the Western World is a respected altruistic leader. However, because Western society is hierarchal, leadership isn't traditionally a mutually accepted role between leader and follower. Thus, leaders can be wonderful or silently disrespected or even hated. Leaders may be incompetent dupes chosen by nepotism or because they are easily manipulated or they may be paid back political appointees. An example of nepotism is a company's hard working owner giving his wife's lazy brother a cushy job.

Nepotism causes other employees to feel unequally treated and sense they are ostracized.

Favoritism and nepotism are common in hierarchal societies, but not in immediate-return societies. Whether prestige is gained by favoritism or hard work is irrelevant in intermediate-return societies since they are unranked. Human foragers "do not value prestige (i.e. a high reputation arising from success of some kind)," Power says. They value most "peace between individuals and groups, autonomy, self-esteem and generalized attachment to social, as well as actual, kin." Any prestige the lazy brother received was underserved like the undue respect given to a dominant ape's unranked brother. Another alpha, the Emperor Napoleon Bonaparte, provided his unranked brothers considerable prestige.

An unranked ape such as Faben without his alpha brother Figan would be subordinate like Evered and Lady Chatterley's lover, two primates relegated to lower rank. Yet, no rank can have privilege exemplified by the lazy brother secure in his job might benefit from his brother-in-law's prestige. Subordinates are not as impressive as dominating individuals like Homicidal Hitler and Horrendous Humphrey. These ruthless primates cloud "our understanding of the balance of power that exists in egalitarian societies," Power says. In ranked societies the hireling, Lady Chatterley's lover, and the subordinate Jomeo, outwardly appear "connected with low rank and submissive behavior."

Crippled Fagan was respected because of his alpha brother, Figan

Joseph Bonaparte
was appointed
King of Spain

by his brother
Napoleon

Leaders and followers in the adaptive egalitarian society have respected-deferred roles without connoting superiority or inferiority assumed in directly competitive settings. The principles supporting egalitarianism would prevent primates like Mike the Kerosene King or Pol Pot the Oriental Oppressor from becoming dictatorial. In an immediate-return system the respected-deferred relationship between charismatic and average chimpanzees or humans is relaxed. Conversely, in the Western World the relationship between a company CEO and a worker on the assembly line would hardly be relaxed or even exist. The chief executive and a worker would unlikely know each other or have reason to exchange roles.

The inequality between capitalists and workers developed with the system that Karl Marx wanted overthrown by force. Earlier, Jean Rousseau wanted to equalize the ranked social organization by removing class distinctions. The two men hoped in vain to reestablish egalitarian principles and emphasize the sustaining properties of the adaptive society. The principles and properties were adapted to operate optimally in an environment in which they were selected. The two intellectual men didn't know the history of the egalitarian society. They couldn't know when its principles and sustaining properties are vacated that disorder fills the void.

Marx and Rousseau never heard of the uncanny similarity of chimpanzee and human behavior and genotypes. There was no evidence in their day that Homo sapiens had inadvertently corrupted their adaptive social organization. Sealing the hopelessness of it all, they didn't know bipedalism trailed by the big brain made a return to yesteryear impossible. The man was yet to be born who said, "You can't go home again."

There is a vast amount of scientific data that wasn't available to help Marx and Rousseau analyze the unsolvable dilemma of human circumstances. Indeed, how discouraging for the two intellectuals or any humanitarian to discover that class distinctions and warfare are innate in a society structured on direct competition and restricted movement with elements of insecurity and fear. Another bit of dispiriting data is many scientists theorize that the only purpose of the creator of life, natural selection, is determination.

Life's creator functions automatically like a physical law to produce the beautiful and ugly. Worst is the depressing thought that survival is temporary and the fact that when life ends it has never been revitalized, except in the imagination. Maybe if the man who said you can't go home again believed there is no awakening he would add you can't go anywhere and it's not worth the struggle for a paradise lost.

BIBLIOGRAPHY

Allen, Bridges, Lyon, Moses and Russell. 1990. *Biology of Mammalian Germ Cell Mutagenesis.* Cold Spring Harbor Laboratory Press

Angela, Alberto & Angela. 1993. *The Extraordinary Story of Human Origins.*Prometheus Books.

Armstrong, Este & Falk, Dean. 1982. *Primate Brain Evolution. Methods and Concepts.* Plenum Press.

Asimov, Isaac. 1987. *Beginnings.* Walker and Company.

Bailey, C. 1948. *The Frontal Lobes.* Comparative Primate Biology Vol. 4. Neuroscience. Liss, Inc.

Beerbower, James. 1968. *Search for the Past.* Prentice-Hall.

Bertell, Rosalie. 1985. *No Immediate Danger.* The Book Publishing Company.

Broom, Robert. 1993. The *Coming of Man. Was it accident or design?* H. F. & G. Witherby.

Chadwick, D. H. 1993. *The Fate of the Elephants.* Viking, London.

Chiarelli, A.B. 1969. *Comparative Genetics in Monkeys, Apes and Man.*Academic Press.

Coon, Carleton. 1965. *The Living Races of Man.* Alfred Knopf.

Crick, Francis. 1981. *Life Itself Its Origin and Nature.* Simon and Schuster.

Darwin, Charles. 1839. Th*e Voyage of the Beagle.*

____.1859. *The Origin of the Species.*

___.1871. *The Descent of Man.* Harvard Classics.

Davis, William. 1993. *The World of Biology.* Holt, Rinehart and Winston.

Dawkins, Richard. 1976. *The Selfish Gene.* 2006. *The God Delusion.*Oxford University Press.. Houghton Mifflin Harcourt.

Delson, Tattersall, Van Couvering and Brooks. 2000. *Encyclopedia of Human Evolution and Prehistory.* Garland Publishing Inc.

De Wall, Franz. 1982. *Chimpanzee Politics: Power and Sex Among Apes.* Harper & Row.

___.1989. *Peacemaking among Primates.* Harvard University Press.

Diamond, Jared. 1992. *The Third Chimpanzee.* Harper Collins. *Dictionary of Scientific and Technical Terms.* 1984. Third Edition. McGraw-Hill.

Encyclopedia of Animals. 1993. Reader's Digest.

Falk, Dean. 1992. *Brain Dance.* Henry Holt and Company.

Feingold & Pashayan. 1983. *Genetics and Birth Defects in Clinical Practice.* Little Brown and Co.

Fossey, Dian. 1983. *Gorillas in the Mist.* Houghten Miffin.

Goodall, Jane. 1989. *Chimps.* Atheneum.

___ 1971 *In the Shadow of Man.* Houghton Mifflin Co.

___ 1967 *My Friends The Wild Chimpanzees.* The National Geographic Society.

___.1986 *The Chimpanzees of Gombe. Patterns of Behavior.* Harvard university Press.

___.1990 *Through a Window.* Houghton Mifflin Co.

Ghiglier, Michael. 1994. *The Chimpanzees of Kibale Forest.* Columbia University Press.

Gould, Stephen. 1997. *An Urchin in a Storm.* Norton.

___.1991 *Bully for Brontosaurus.* Norton.

___.1995 *Dinosaur in a Haystack.* Harmony Books.

___.1993 *Eight Little Piggies.* Norton.

___.1977 *Ever Since Darwin.*

___.1985 *The Flamingo's Smile.* Norton.

___.1983 *Hen's Teeth and Horses Toes.* Norton.

___1980 *The Panda's Thumb.* Norton.

___.1987 *Time's Arrow.* Harvard University Press.

Hoobler, Thomas Dorthy. 1993. *Confucianism World Religions.* Facts on File, Inc.

Johanson, Donald and Lenora. Edga, Blake. 1994. *Ancestors.* Random House.

Johanson, Donald & Blake, Edgar. 1996. *From Lucy to Language.* Simon and Schuster.

Johanson, Donald and Edey, Maitland. 1981. *Lucy.* Simon & Schuster.

Johanson, Donald & Shreevei, James. 1989. *Lucy's Child.* William Morrow & Company.

Kalter, Harold. 1983. *Issues and Reviews in Teratology.* Plenum Press.

King, Max. 1993. *Species Evolution, the role of chromosome change.* Cambridge University Press. Cambridge, MA.

Kratochvil, Clyde. 1972. *Chimpanzee: Immunological Specificities of Blood.* S. Karga.

Lawrence, D.H. 1928. *Lady Chatterley's Lover.* Greenwhich House.

Leakey, Richard. 1981. *The Making of Mankind.* E.P. Dutton. New York.

___.1994 *The Origin of Humankind.* Harper–Collins Publishers.

___.1982 *Human Origins.* E.P. Dutton, Inc.

___.1992 *Origins Reconsidered.* Doubleday

Marcus, Eric. *Is it a choice?* Harper Collins.

Margulis, Lynn. 1992. *The Diversity of Life. The Five Kingdoms.* EnslowPublishers, Inc.

Marx, Karl. 1848 *The Communist Manifesto.*

___.1872 *Das Kapital.*

Mayr, Ernst. 2001. *What evolution IS.* Basic Books.

Metalious, Grace. 1956. *Peyton Place.* Julian Messner.

Mead, Margaret. 1928. *Coming of Age in Samoa.* William Morrow & Company Company.

Mill, John Stuart. 1848. *Principles of Political Economy.*

Milner, Richard. 1990. *The Encyclopedia of Evolution.* Owlet.

Milton, John. 1667. *Paradise Lost.*

Mitchell, Margaret. 1936. *Gone with the Wind.* Macmillian Publishing Co.

Modern Biology. 1989. Holt, Rinehart and Winston.

Morrel, Virginia. 1995. *Ancestral Passions. The Leakey Family and the Quest for Humankind's Beginnings.* Simon and Schuster.

Morris, Desmond & Ramona. 1966. *Men and Apes*. McGraw–Hill Book Company.

Morris, Desmond. 1967. *The Naked Ape*. McGraw–Hill.

Nanny, David. *Evolutionary Discontinuities*. Personal Website.

Nishida, Toshisada. 1990. *The Chimpanzees of the Mahale Mountain. Sexual and Life history Strategies*. University of Tokyo Press.

Pilbeam, David. 1972. *The Ascent of Man*. The Macmillian Company.

Power, Margaret. 1991. *The Egalitarians–Human and Chimpanzee*. Cambridge University Press.

Ridley, Matt. 1993. *The Red Queen*. Penguin Books.

Rousseau, Jean–Jacques. 1775. *Discourse on the Origin of Equality*.

____.1762. *The Social Contact*.

Sagan, Carl & Druyan, Ann. 1992. *Shadows of Forgotten Ancestors*. Random House.

Scroggs, Robin. 1983. *The New Testament and Homosexuality*. Fortress Press.

Shakespeare, William. *Hamlet, Henry IV, Julius Caesar, Romeo and Juliet*.

Smith, Adam. 1776. *The Wealth of Nations*.

Standard History of the World. 1931. Standard Historical Society.

Strum, Shirley. 1987. *Almost Human*. Random House.

Stuart, King James VI. 1611. *The Holy Bible, King James Version*.

Wallace, Bruce. 1966. *Chromosomes, Giant Molecules, and Evolution.* W.W. Norton & Co. Inc.

Watson, James. 2003. *DNA The Secret of Life.* Alfred A. Knopf.

Wrangham, Richard. 1994. *Chimpanzee Cultures.* Harvard University Press.

Wynbrant, James & Ludmas, Mark. 1991. *The Encyclopedia of Genetic Disorders and Birth Defects.* Facts on File, Inc.

INDEX

GLOSSARY

Adaptation or adaption-- A physical or behavioral trait of an organism that adds to its fitness.

Adaptive radiation-- New species moving into and adjusting to various niches.

Ad hoc adaptions-- Traits selected for the moment.

Advanced-- Poor term for organisms with intricate life systems.

Agricultural Revolution-- A transition 10,000 years ago when hunter-gatherer tribes started farming and competing directly for resources.

Allele-- A form of a gene that is a unit of inheritance.

Altruism-- Helping others, theorized in part to promote transmission of the selfish gene.

Amino acids-- Molecules called the building blocks of life that construct larger molecules of proteins.

Anthropomorphism-- Attributing human characteristics to animals.

Anthropology-- The study of human societies.

Ape-- A primate with no tail.

Artificial selection-- Breeding animals and plants with preferred traits for human use.

Asexual Reproduction-- Producing offspring without uniting a sperm and ovum.

Australopithecine-- A genus of early hominids with ape sized brains that walked upright suggestive of chimps on human legs.

Aves-- The animal class of birds.

Background extinction-- The steady extinction of a number of species throughout geological history.

Backtrack-- False belief evolution can be reversed.

Baldwin effect-- Selecting traits that strengthen a body part or behavior.

Baldwinized-- Genes ingrained into the genotype by the Baldwin effect.

Basicranium-- The skull floor.

Bipeds-- Animals that walk upright on two legs, usually applied to hominids.

Biological determinism-- The belief that genes and not culture determines human behavior.

Blood network-- The arrangement of blood vessels in the brain.

Bonobo-- Chimpanzee subspecies and third closest human relative.

Carnival-- Jovial Meeting of different human hunter-gatherer tribes or different chimpanzee communities.

Characteristic or trait-- A physical or behavioral attribute such as eye color or social cooperation resulting from a gene or genes and influenced by the environment.

Chromosome-- A unit that stores genes.

Clone-- Genetically identical organism produced by asexual reproduction such as bacteria.

Coevolution-- Evolution of two interdependent organisms that at least one depends on the other such as bees and flowers.

Coding genes—Genes that produce a physical or behavioral trait.

Co-dominant genes-- Genes that exert equal influence on a trait such as skin color.

Co-opted genes-- Existing genes used by natural selection.

Common ancestor-- The progenitor of descendent species such as the ape that gave rise to Australopithecines and humans.

Common descent Theory-- Poses that related organisms descended from a common ancestor such as Homo habilis gave rise to Homo erectus and Homo sapiens.

Conspecifics-- Members of the same species.

Constraints-- Keep evolution from occurring.

Coincidental adaptations-- Different forms of a trait selected for the same purpose.

Convergent evolution-- Adapting similar structures by unrelated species in a similar environment.

Consortship-- Chimpanzee honeymoon.

Cro-Magnon-- Homo sapiens that coexisted with Neanderthals.

Darwinism-- Evolutionary modes proposed by Charles Darwin such as natural selection and gradualism.

Discontinuity-- Space or gap in the fossil record or between categories of organisms.

Dead-end-- An ambiguity that the last organism in an evolutionary series such as humans will remain last if they don't evolve into a new species.

Degenerate-- Poor description for a primitive or descendent animal such as a frog that descended from an earlier animal class.

Dominant gene-- Overrides the influence of other genes for traits such as brown eyes are dominant over blue eyes.

DNA-- Genetic recipe to construct an organism during reproduction.

Duckbill platypus-- Intuitively the odd animal or degenerate with traits of mammals and reptiles.

Elevator Hypotheses-- Proffers animals adjust their vascular systems to offset the effects of gravity.

Encephalization-- Brain size increase during evolution.

Egalitarian society-- Adaptive human and chimpanzee social organization wherein all members are equal.

Embryo-- Post zygote that develops into a fetus paced by regulatory genes.

Equilibrium-- Constant gene frequency in a gene pool.

Evolution-- Change in organisms over time.

Exaptation-- A part of a precursor that was modified for use in a descendant such as an anterior leg into a wing.

Expensive Tissue Hypothesis-- Poses a part requiring much energy acquires it at the expense of another part such as the large human brain accounts for the small intestinal tract.

Favorable genes-- Genes producing the best traits for natural selection.

Fission and fusion-- Individuals freely banding and disbanding within a social group.

Flagella-- A threadlike appendage on some unicellular organisms used for locomotion and considered the precursor of sperm tails.

Founder effect-- Rapid evolution in a small group that migrated from a larger population.

Founders-- Rapidly evolving members of a small group of emigrants from a larger population.

Fundamentalism-- Zealous conformance to puritanical doctrine.

Fossils-- Remains of an organism that are usually teeth and bones.

Foramen magnum-- Skull opening for the spinal cord.

Fungi or funguses-- Organisms such as mushrooms that digest decaying matter.

Gene-- A transmissible unit stored on a chromosome that determines a physical or behavioral trait.

Gene pool-- Total genes in a population.

Genetic code-- The biological basic of heredity.

Genetic drift-- Gene frequency in a small isolated population that initiates evolution without natural selection.

Genetic turnover-- New and old genes entering and exiting a population throughout generations.

Genetic variation-- Genes available for natural selection to choose.

Genotype-- Total number of an organism's genes.

Genus-- Next classification above species such as the genus Homo.

Gestation-- Period an offspring develops during pregnancy.

Gombe National Park, Africa--Location of Jane Goodall's chimpanzee study.

Gorilla-- A Great Ape and third closest human relative.

Gracial Australopithecine-- Early hominid that moved to the hot African plains and theoretically evolved big brains.

Gradualism--Theory proposing that evolution occurs slowly.

Group selection-- Theory holding social groups evolve by natural selection.

Holocaust or genocide-- A survival adaption of humans and chimpanzees to eliminate competitors and rationalized by extreme pseudospeciation.

Homo lineage-- Evolutionary line of hominids ending with humans.

Homo sapiens-- Humans sometimes classified as Great Apes.

Homosexuality-bisexuality Hypothesis-- Proposes human anomalous sexuality is a product of a lopsided brain.

Hominids-- Humans and their extinct bipedal relatives that arose in Africa.

Hierarchy-- Nonadaptive ranking order of a corrupted equalitarian society.

Hybrid-- An offspring of two different species.

Hunting-gathering-- Adaptive human and chimpanzee method of food acquisition.

Heaven-- A survival mechanism and a promised payoff for obeying social rules.

Hell-- A frightening invention to insure obedience of theocratic decrees.

Hybrid--Offspring of two different species

Hypothesis-- An assumption or educated guess based on observation and some facts.

Hxaro-- Exchanging gifts with little or no value to promote friendliness.

Individual differences-- Physical and behavioral variations among members of a species.

Intelligent design-- Belief evolution is a planned.

Intermediate animal-- An animal between its precursor and descendant.

Junk genes-- Noncoding genes with no apparent purpose.

Juvenile human endocrine system-- The subadaptive hormonal glands of humans.

Karyotype-- Usually a picture of the chromosomes of a species or an individual including number, form, and size of the chromosomes.

Knuckle walkers-- Apes that use their knuckles for support when walking.

Lamarckism-- False belief acquired characteristics are inherited.

Living fossil-- An animal existing long after most of its species became extinct.

Lopsided-- Describes disproportionate shape of human brain.

Love-- Broad term for the positive emotions that sustain an egalitarian society.

Lucy--- Partial fossil skeleton of a female Australopithecine

Macroevolution-- Evolution above the species level such as reptiles evolving into mammals.

Microevolution-- Evolution at the species level such as chimpanzees giving rise to bonobos.

Maladaptive-- Describes a feature of an organism that seems inefficient.

Mammal-- Animal with milk glands and hair such as humans and dogs.

Marriage-- A nonadaptive institution.

Marsupial-- Nonplacental subclass of mammals

Meiosis-- In sexual organs a process that divides chromosomes and mixes genes to insure individual differences among offspring.

Missing link-- Antiquated term for a theoretical primate between the ancestral ape and humans.

Mitosis-- A growth process of dividing body cells with an equal number of chromosomes.

Mosaic Evolution-- Natural selection favoring specific body parts.

Mule-- The hybrid between a horse and donkey.

Multiregional Theory-- Proposes humans evolved concurrently worldwide.

Murder, mayhem and suicide-- Consequences of corrupting the adaptive human and chimpanzee society of egalitarianism.

Natural selection-- Theory that in every generation the weakest members of a population are eliminated and the strongest members survive.

Nature-- Term interchangeable with natural selection.

Negative emotions-- Responses such as hate and jealousy that break social bonds.

Neoteny—The retention of juvenile traits in the adults of a species.

New genes-- Recently mutated genes recruited by natural selection.

Niche-- An area conducive for adaption such as the underground niche of moles.

Nonadaptive-- A substandard part or corrupted condition such as a directly competitive human society altered from an indirectly competitive egalitarian society.

Noncoding genes-- Genes that don't synthesize a protein to produce a trait.

Nonselective sexuality-- Multiple sexual partners with no special preference.

Normalizing or stabilizing selection-- Eliminating extremes to maintain an average phenotype or behavioral characteristic.

Nucleotide-- A chemical compound and component of DNA.

Old genes-- Genes that were present in preceding organisms.

Omission Hypotheses-- States the human brain is not a product of natural selection and questions its adaptiveness.

Omnivore-- Organisms that eat meat and plants such as bears, chimps and people.

Organism-- Any life form.

Orangutan-- A Great Ape and fourth closest human relative.

Ostracism-- Method adapted by the original human and chimpanzee egalitarian society to control deviants.

Ostrich-- Largest terrestrial bird whose ancestors were adapted for flight.

Out of Africa Theory-- Poses Homo sapiens evolved suddenly in Africa from an unspecified species two or three hundred thousand years ago.

Ovum-- Female reproductive cell or egg with half number of chromosomes.

Paleoanthropology-- The study of prehistoric human societies.

Paleoneurology-- The study of extinct hominid brains.

Paramammal-- Extinct animal with mixed traits of mammals and reptiles.

Phenotype-- The physical body and behavioral characteristics of an organism.

Pili-- Tiny appendages on small organisms thought to be precursors of flagella.

Plate tectonics-- Theory that the earth's crust consists of slowly moving plates.

Plates-- The composite parts of the earth's crust.

Polypeptide-- Molecule built by amino acids.

Population-- An interbreeding number of conspecific animals wherein most evolution occurs.

Positive emotions-- Responses such as love and compassion that strengthen social bonds.

Preadaptation-- A body part or behavior present in a species that intelligent designers believe are to be used in the descendants.

Prefrontal cortex-- Thinking area of brain that is disproportionately large in humans.

Primate-- Human, ape and monkey.

Primitive-- Poor term for organisms with the fewest life systems.

Progression-- Belief evolution moves in a line from primitive to advanced organisms.

Pseudo-extinction-- A species evolving into a new species instead of becoming extinct.

Pseudospeciation-- Members of a social animal such as humans and chimpanzees seeing other members of the society as a different species.

Prime mover-- Causes evolution to occur.

Prime releaser-- Allows evolution to occur.

Punctuated equilibrium-- Theory posing that species form rapidly.

Quadruped-- An animal that walks on all fours, usually applied to primates.

Radiator Hypothesis-- Proposes a brain will grow only as large as its blood network can prevent the brain from overheating.

Recapitulation-- Theory posing embryos repeat stages during gestation of their precursors.

Regulatory genes-- Speed up or slow down embryonic development.

Reptile-- Cold blooded scaly vertebrate that usually lays eggs such as snakes.

Robust Australopithecine-- Vegan with large jaws and last of the genus to meet extinction.

Saltation-- Rapid speciation by a mutation.

Selective pressure-- Convenient, but erroneous term that threatened organisms are pressured to evolve.

Selective significance-- A measure of the quality of an adaption.

Selfish gene-- Theory that genes temporarily use the bodies that house them to transfer from generation to generation.

Sexual selection-- Choosing a mate with the most attractive traits.

Sibling species-- A related species whose members are nearly identical but can't produce fertile hybrids.

Slavery-- The practice of forced servitude on people and animals initiated during the Argricultural Revolution by the corruption of the human adaptive social organization.

Social norm-- Standard of behavior conformed to by individuals of a society.

Speciation-- The formation of a new species from its precursor species.

Species-- Organisms that look alike and are almost genetically identical except for a few genes that produce individual differences.

Sperm-- Male reproductive cell with a half number of chromosomes.

Sponge-- A primitive dead-end animal.

Stasis or static-- Inactive evolution in a species.

Subadaptive-- Intuitively, a biological system or unit that functions inadequately.

Subspecies-- A species subdivision that can interbreed and is usually isolated.

Superadaptive-- Intuitively, a biological system or unit that functions superbly.

Synthesis-- All fields of evolutionary research combined to explain evolution.

Teleological-- Belief in a purpose of natural phenomena.

Terrestrial-- Living on land.

Tetrapod-- Animal with four legs such as a lizard.

Theory—An explanatory statement with more facts than a hypothesis.

Transitional animal-- An intermediate species that is becoming a new species.

Turkana boy-- An extinct hominid with an estimated brain size one-third smaller than the human brain.

Utopia-- Symbol of original human or chimpanzee society of egalitarianism.

Vestige-- A remnant structure of an organ or body part from an earlier stage of evolution.

Virus-- An organism without all the life processes.

Voyeurism-- Adaptive reproductive method of chimpanzees and humans to become sexually stimulated by watching couples copulating.

War-- A result of the corrupted egalitarian society of humans and chimpanzees.

Western World—Symbolizes a materialistic society with uninhibited sexual practices.

Xenophobia-- Fear of strangers

X-rays-- Particles that penetrate a reproductive organ causing a mutation by altering a nucleotide.

Zygote—A united male and female sex cell with a complete number of chromosomes.

Years Ago	Epoch	Period/Age		Era	Eon	Major Events
Present day	Holocene					End of ice age and rise of modern civilization
11,430	Pleistocene	Neogene				Extinction of many large mammals. Evolution of fully modern humans
1.81 million	Pliocene			Cenozoic		
5.33 million	Miocene					
23.0 million	Oligocene					
37.2 million	Eocene	Paleogene				Appearance of first "modern" mammals
55.8 million	Paleocene				Phanerozoic	
65.5 million		Cretaceous				Dinosaurs reach peak, become extinct. Primitive placental mammals
146 million		Jurassic		Mesozoic		Marsupial mammals, first birds, first flowering plants
200 million		Triassic				First dinosaurs, Egg-laying mammals, breakup of Pangea into Gondwana and Laurasia
251 million		Permian		Paleozoic		Permian extinction event- 95% of life on Earth becomes extinct
299 million		Carboniferous[1]	Pennsylvanian			Abundant insects, first reptiles, coal forests
318 million			Mississippian			Large primitive trees, first land vertebrates

Age	Period	Era	Eon	Events
359 million	Devonian			First amphibians, clubmosses and horsetails appear, progymnosperms (first seed bearing plants) appear
416 million	Silurian			First vascular land plants, first jawed fish
443 million	Ordovician			Invertebrates dominant; first land plants
488 million	Cambrian			Major diversification of life in the Cambrian explosion
542 million	Ediacaran			First multi-celled animals
630 million	Cryogenian	Neoproterozoic		Possible snowball Earth period, Rodinia begins to break up
850 million	Tonian		Proterozoic[2]	First acritarch radiation
1.0 billion	Stennian			Formation of Rodinia
1.2×10^9	Ectasian	Mesoproterozoic		
1.4×10^9	Calymmian			
1.6×10^9	Statherian	Paleoproterozoic		First complex single-celled life
1.8×10^9	Orosirian			Transition to oxygen atmosphere
2.05×10^9	Rhyacian			
2.3×10^9	Siderian			
2.5×10^9		Neoarchean		
2.8×10^9		Mesoarchean		
3.2×10^9		Paleoarchean	Archaean[2]	
3.6×10^9		Eoarchean		Simple single-celled life
3.8×10^9			Hadean[2,6]	4.1×10^9 - Oldest known rock; 4.4×10^9 - Oldest known mineral; 4.57×10^9 - Formation of Earth